THE  TIMES
Night SKY COMPANION

Times Books
77–85 Fulham Palace Road
London W6 8JB

This edition first published by Times Books 1999
© Michael Hendrie 1999

The Times® is a registered trademark of Times Newspapers Ltd.,
a subsidiary of News International Plc.

The HarperCollins website address is:
www.**fire**and**water**.com

ISBN 0 7230 1041 2

10 9 8 7 6 5 4 3 2 1
04 03 02 01 00 99

Designed by Colin Brown
Colour reproduction by Colourscan, Singapore
Printed and bound in Hong Kong by Printing Express Ltd.

Michael Hendrie has been writing the monthly Night Sky column in *The Times*
and annual Night Sky booklet since 1989. He has been an active observer since
1950, specialising in the photography of comets and the Sun. He joined the British
Interplanetary Society and British Astronomical Association in 1951, and was
elected a fellow of the Royal Astronomical Society in 1956. He was Director of the
BAA Comet Section from 1977 to 1986 and was awarded the Association's Walter
Goodacre Medal. The International Astronomical Union named asteroid 4506
Hendrie in recognition of his services to astronomy. He has lectured to adult edu-
cation classes, astronomical societies and other groups and has been the author of
many astronomical papers, articles and reviews. Since retirement from an international oil company, he has been able to devote more time to making and using his
own instruments and writing about astronomy.

THE TIMES
Night SKY
COMPANION

Michael J. Hendrie

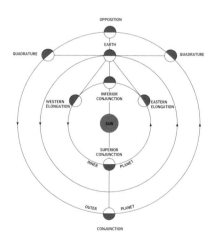

CONTENTS

Pluto

PREFACE

This Companion is intended to supplement the monthly astronomical notes in *The Times* and the annual booklet *The Times Night Sky*. The monthly notes were first published in *The Times* in 1919. I first discovered the booklet in 1952. In 1966 *The Times Guide to the Sky at Night* by Norman Wymer was published explaining, in simple terms and with diagrams, many of the basic phenomena that can be observed in the night sky with the naked eye. The Forward was by *The Times* Astronomical Correspondent, Dr W H. Steavenson.

The annual Night Sky booklet is intended to provide a simple guide to what can be seen with the naked eye from the latitudes of the British Isles during the forthcoming year. It shows in two polar diagrams the positions of the planets in their orbits and in a further twelve monthly charts, those stars, planets and the Moon above the horizon in the late evening. Opposite each chart are notes on all the planets, their brightness and visibility, any close approaches to the Moon and other planets, the phases of the Moon and times of the equinoxes and solstices. Notes on the visibility of eclipses are given. Prospects for the year's major meteor showers and unusual phenomena are included in the booklet under The Year in Space, as well as an outline of some important astronomical discoveries and space missions.

The Times Night Sky is published in November for the following year. The monthly notes in *The Times* are written only days before publication so they can include not only the appropriate monthly chart from the booklet and notes on the Moon and planets, meteors, sunrise, sunset, twilight times and eclipses, but also any unusual events in the next few weeks such as bright comets and novae which may not have been discovered when the booklet was finalised. There is also room in the notes for an explanation of some current event taking place that month, such as an eclipse or close grouping of some planets. In the absence of anything unusual, brief explanations of other night sky phenomena are given.

What neither publication can include are more detailed explanations with supporting diagrams and illustrations of how the various phenomena arise, tabular data on the stars and planets, an explanation of terms used in astronomy and simple advice on observing the night sky. It would not be practicable to repeat this type of information each month or even every year. It can best be contained in a separate companion volume that can be used alongside the monthly notes and annual booklet by those readers who wish to know more about what they are observing.

It is the purpose of this Companion to increase the observer's enjoyment of the night sky by offering further explanation of what can be seen with the naked eye and the occasional use of ordinary binoculars. Some events are not very frequent and bad weather and other commitments reduce opportunities to see them. If you get the chance to see something interesting do not miss it: you may have to wait a long time for another opportunity.

For example, in looking for illustrations for this guide, I found my only complete set of total lunar eclipse photographs was taken as long ago as 1954; I had always intended to get a better set next time but never have. Photographs of Sputnik 2 and Apollo 8 cannot be taken again. It has been possible to fill a few gaps with recent pictures but in astronomy one cannot make things happen, one can only wait until they do. So the old records have been valuable. I am fortunate in being able to use some of the excellent illustrations by Richard McKim and the late Harold Ridley.

I must thank John Vetterlein for reading the manuscript and making valuable suggestions about the wording of some sections of the text. Any ambiguities that remain are my own responsibility.

Michael J. Hendrie

CELESTIAL SPHERE

Quite simple rules can be given on what is visible in the night sky and simple explanations of how the different phenomena such as planetary movements and eclipses arise. It is the purpose of the *Night Sky Companion* to provide these. To avoid misunderstandings about the accuracy of these rules and explanations it should be made clear that the principles are correctly stated here in enough detail for the naked eye observer, but there are many more small but important factors that have been omitted. The Universe is not simple and much of what it contains cannot be explained in simple terms. Also, although there are cycles of events, nothing ever repeats itself exactly. Numbers have been rounded to a decimal place or two and the prefixes 'about' or 'approximately' generally omitted from many statements.

Readers who want to delve more deeply into any aspect of what is covered in this Companion can find books of any degree of complexity, up to a full mathematical treatment. But this is not necessary for a general understanding of the phenomena discussed and for enjoyment of the night sky with the naked eye.

DISTANCES TO THE STARS

From our position on the surface of the Earth the stars appear to lie on the inside of a spherical surface. We get no impression of the distance to the stars by just looking at them. They could all be at the same distance just differing in brightness, as early observers used to believe, or they could extend in depth like trees in a forest.

When astronomers first tried to measure the stars' distances they assumed, as a starting point, that the brightest stars would be the nearest. This turned out to be a poor guide as stars differ greatly in their intrinsic brightness. Faint stars are much more common among the nearer stars as well as in other parts of the Galaxy. Our familiar constellations are represented by a high proportion of brighter but less common stars that can be seen at greater distances.

The stars are so far away that distances in kilometres or miles mean very little and lead to inconveniently large numbers. Astronomers use the parsec, which is the distance of a star at which the radius of the Earth's orbit round the Sun (1 AU) subtends an angle of 1 arcsecond. The distance of such a star 1 parsec away is also equal to 206265 Astronomical Units. A more easily understood unit is the light year (3.26 ly = 1 parsec), the distance travelled by a pulse of light or radio waves in one year.

Light travels at 300,000 km (186,000 miles) per second which is about 9.5 million million km per year and the nearest star Alpha Centauri is 4.3 ly or 41 million million km away. The most distant named star (Deneb in Cygnus) shown on The Times star charts is over 1800 ly away and the most distant object shown, the spiral galaxy in Andromeda (M31), over 2 million light years. There is little point in converting these great distances into kilometres.

CELESTIAL HEMISPHERE

Although the stars are moving relative to each other, this is not discernible to the naked eye observer in less than centuries and so the stars can be considered for convenience as fixed to the inner surface of the arbitrarily large celestial sphere. From a point in space away from the Earth we should see stars all around us, but from the surface of the Earth only half of the sky is visible at any one time, the Earth beneath our feet hiding the other half of the sky from view.

What we see at night is a hemisphere of stars: while we shall still see a hemisphere from different places on the Earth the stars it will contain will be different. They will also be different depending on the season and time of observation.

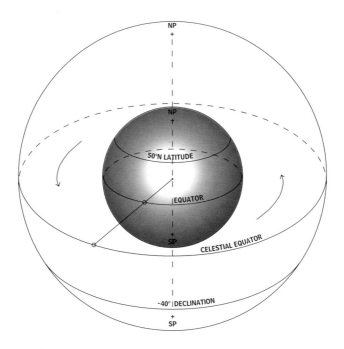

14   Fig. 1a   *Celestial sphere and rotating Earth*

THE ROTATING EARTH

In Figure 1a we see the celestial sphere surrounding the Earth. The celestial sphere is divided into two equal halves or hemispheres by the celestial equator. This lies immediately above the Earth's equator so that a line from the centre of the Earth passing through the Earth's equator also passes through the celestial equator. Another way of saying this is that the celestial equator passes through the zenith (immediately overhead) of all points on the Earth's equator.

The imaginary Earth's axis, about which it rotates once each day, meets the surface of the Earth at two points, the north and south geographical poles. Extending this axis further in both directions makes it cut the celestial sphere at the north and south celestial poles. Again one may say that the celestial poles are at the zenith for observers at the geographical poles.

The stars appear to move across our sky from east to west each night, the celestial sphere turning about the two celestial poles, but in reality the directions to the stars are fixed and it is the Earth that rotates in the opposite direction, anticlockwise as seen from above the north pole.

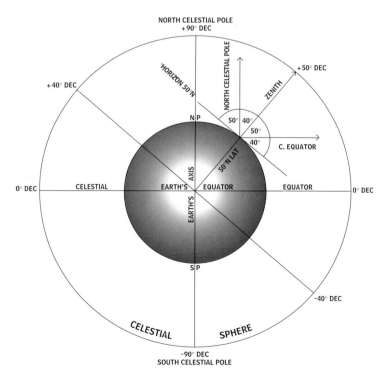

Fig. 1b *Relation between declination and latitude for 50° north latitude*

LATITUDE AND DECLINATION

In Figure 1b we see the relation between latitude on the Earth and declination on the sky. Latitude is measured from the equator at 0 degrees to the north pole at 90 degrees N and south pole at 90 degrees S. Declination is measured from the celestial equator at 0 degrees to the north and south celestial poles though for convenience in calculations these are usually expressed as +90 and –90 degrees.

Declination like latitude and longitude is measured in degrees (°), minutes (') and seconds (") of arc (angle), there being 60" in a minute, 60' in a degree and therefore 3600" in a degree. We do not need this accuracy here though.

Figure 1b shows two examples of declinations, +50 degrees and –40 degrees. The significance of these for us is explained in the figure. The British Isles is situated between latitudes 50 degrees N and 61 degrees N. London is at 51.5 degrees N. For simplicity one latitude has to be chosen for the diagrams and there are advantages later in making this the most southerly, 50 degrees N.

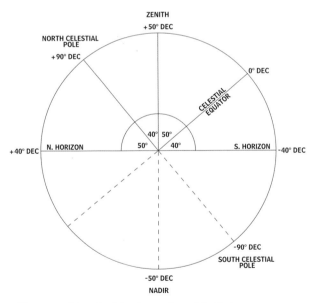

Fig. 1c *Elevation of the celestial pole at 50° north latitude*

HORIZON, ZENITH AND CELESTIAL POLE

Assuming the Earth to be a sphere (it is a globe slightly flattened towards the poles called an oblate spheroid), we have drawn the horizon (tangent to the sphere) at 50 degrees N. It is clear that the declination of stars at the zenith will be also be +50 degrees and the north celestial pole will be 40 degrees from the zenith or 50 degrees above the northern horizon as the total angle from zenith to horizon is always 90 degrees.

In Figure 1c we have redrawn this part of Figure 1b by turning it anti-clockwise through 40 degrees so that the zenith is now at the top and the horizon is now horizontal which is how we see it. The nadir is the point opposite in the sky to the zenith, towards the centre of the Earth and immediately below our feet.

ALTITUDE OF THE CELESTIAL POLE ABOVE THE HORIZON
So one rule is that the altitude of the celestial pole above the horizon is equal to the latitude of the place of observation. This can be seen in Figure 1c. Note also that the celestial equator must reach an altitude above the southern horizon of 40 degrees. This means that for a star to be seen from even the most southerly part of the British Isles it must have a dec-lination of no more than −40 degrees. Those stars south of this declina-tion, between −40 degrees and −90 degrees never rise above our southern horizon.

CIRCUMPOLAR AND OTHER STARS
Figure 1c shows another interesting rule. Stars on the northern horizon can be no more than 50 degrees from the pole of the sky, that is their declination cannot be less than +40 degrees. Stars north of +40 degrees are said to be circumpolar at latitude 50 degrees N and never set below the northern horizon. So there are three groups of stars, those with declina-tions from +90 degrees down to +40 degrees which never set, those from +40 degrees to −40 degrees which do set but can be seen sometimes and those south of −40 degrees which can never be seen from 50 degrees north latitude.

(It is not difficult to work out the corresponding figures for other northern latitudes. Similar considerations apply in the southern hemi-sphere.)

STARS VISIBLE FROM THE NORTH POLE, 50 DEGREES NORTH AND THE EQUATOR
Figure 1d shows the situation for an observer in the British Isles (centre), at the north and south poles (bottom) and on the equator (top). Note that from the equator the north and south poles lie on the horizon and all stars are visible at some time or other but there are no circumpolar stars. At the north pole, only those stars in the northern hemisphere are vis-ible and they are all circumpolar stars but the stars in the southern celes-tial hemisphere remain below the horizon throughout the year.

LONGITUDE AND RIGHT ASCENSION
On the Earth we locate places by their latitude and longitude. Parallels of latitude (except for the equator) are small circles, of ever shorter cir-

cumference as the poles are approached. They correspond directly with declination on the sky as we have already seen.

In Figure 1e the circles of longitude are all the same size, passing through both poles and cutting the equator at right angles. On the Earth they are fixed in relation to a chosen place, the zero of longitude passing through the Old Greenwich Observatory in London. On the sky they are fixed in relation to the celestial poles and the intersection between the celestial equator and the ecliptic (see p.42).

These celestial great circles could have been labelled with degrees of longitude but (for good reasons) they are not, longitude having another meaning in astronomy. These circles are labelled with hours, minutes and seconds of Right Ascension. RA as it is generally known is measured in hours (h), minutes (m) and seconds (s) of time (not arc or angle). This may seem perverse but in practice is very helpful.

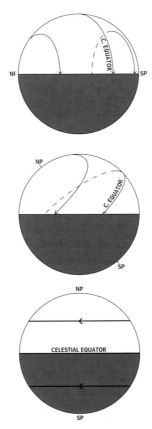

Fig. 1d *Circumpolar and other stars*

RELATIONSHIP BETWEEN UNITS OF ANGLE AND TIME

In Figure 1e we have the celestial sphere with the equator and a typical parallel of declination shown. We also have three great circles of RA at 30 degree intervals as seen by the observer at O on the Earth. Note that they pass through the poles of the sky and cut the equator at right angles. As there are 360 degrees in a circle and 24 hours in a day it follows that 1 hour of RA is equal to 15 degrees at the equator, and our circles of longitude which are 30 degrees apart will also be 2 hours of RA apart.

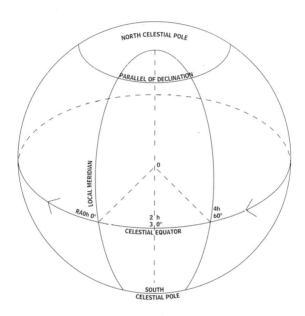

Fig. 1e *Right ascension and the local meridian*

THE LOCAL MERIDIAN

The great circle that passes through the zenith and the north and south points at any place is called the local meridian. The rotation of the Earth will cause an observer at O to see stars moving westwards and successively crossing the local meridian.

THE ZERO POINT OF RIGHT ASCENSION

For the moment we can imagine a single star chosen to represent RA 00h 00m 00s (where the celestial equator and ecliptic cross). Then the moment the star crosses the local meridian (is due south) our clock is set to read 00h 00m 00s (or 0h for short) local time. Two hours later stars which were 30 degrees to the east will be crossing our local meridian and our clock will read 2h local time. These stars are said to have a RA of 2h 00m.

RIGHT ASCENSION AND DECLINATION DEFINE THE POSITIONS OF ALL CELESTIAL OBJECTS

Together the RA and Declination of a star uniquely define its place on the celestial sphere or a chart of the stars as latitude and longitude do on the Earth or on a map of the Earth. Also if we know our local time, we know which stars are due south at that moment and therefore where all the other stars are at that moment and so we can find any star again at any time of the night throughout the year.

CLOCK STARS AND LOCAL TIME

Before the arrival of the telegraph, all observatories kept their own local time, using a small star transit telescope permanently mounted to move only in an arc along the local meridian. The time shown by the observatory clock as each bright clock star crossed the meridian would be noted and then compared with the Right Ascension taken from an accurate star catalogue. From this the error of the clock was easily established. By tabulating the nightly differences the gaining or losing rate of the clock was found and could be used to correct the time shown by the clock at any other instant. The clock's rate could be adjusted to keep the differences conveniently small.

THE STAR CHARTS IN THE TIMES NIGHT SKY

The circles showing RA and Declination are omitted from the monthly charts used in The Times Night Sky as these charts are necessarily small and clarity is important. The celestial equator only is shown. The twelve monthly charts are reproduced in Figures 1f (1–12) with the local meridian (and ecliptic) added. For example about the beginning of January at 23h GMT (the time for which the chart is drawn) the stars lying from north to south down the centre of the chart have a RA of 5.7h or 5h 40m. In the February chart these stars can be found to the west and have been replaced by those that were 30 degrees farther east in early January: they have a RA of 7.7 h. As the sky advances westwards by 2h every month and there are twelve months in a year, the same stars will be on the meridian again at the start of January in the next and subsequent years.

SIDEREAL TIME IS NOT SUITABLE FOR EVERYDAY USE

The moment a star crosses the meridian can be measured very precisely and this time by the stars, called sidereal time, is used by astronomers in their observations and to check our everyday clocks. But our daily lives are governed by the Sun and daylight, not by the positions of the stars, so sidereal time is not suitable as 12h can be midnight in March and midday in September. We need a time which relates to the hours the Sun is above the horizon and, because we are travellers, a time that does not differ by awkward amounts from place to place and which is synchronised worldwide. Solar time is explained on p.45.

Fig. 1f   *The twelve monthly star charts*

1: January

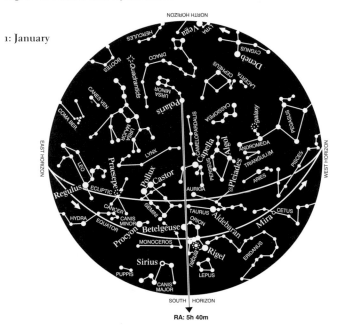

RA: 5h 40m

2: February

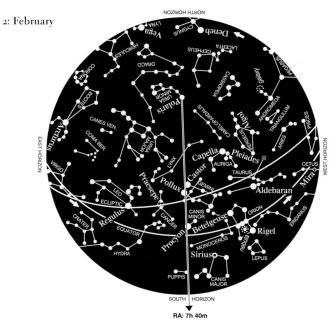

RA: 7h 40m

3: March

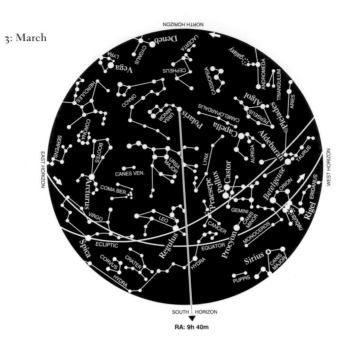

RA: 9h 40m

4: April

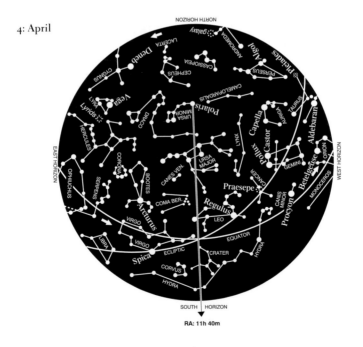

RA: 11h 40m

5: May

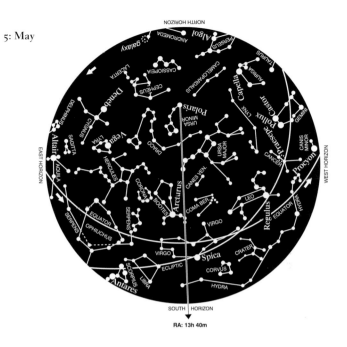

RA: 13h 40m

6: June

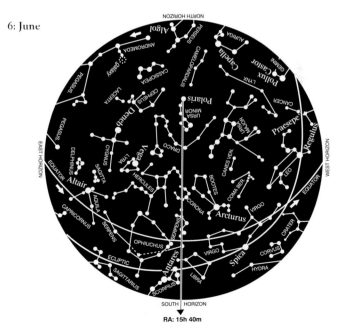

RA: 15h 40m

7: July

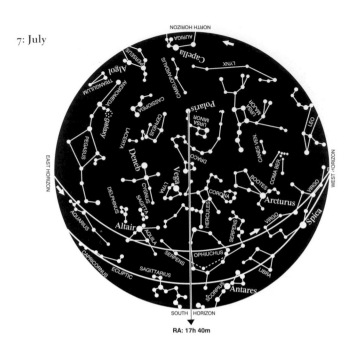

RA: 17h 40m

8: August

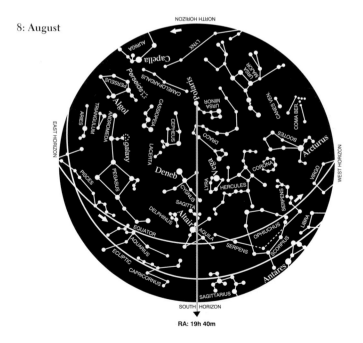

RA: 19h 40m

9: September

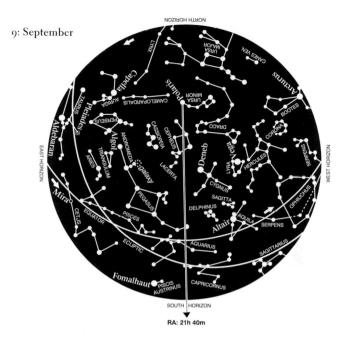

RA: 21h 40m

10: October

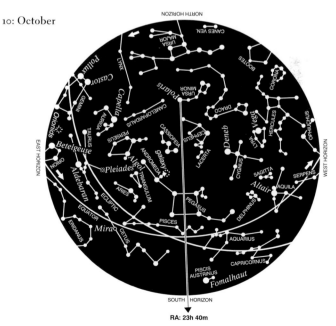

RA: 23h 40m

11: November

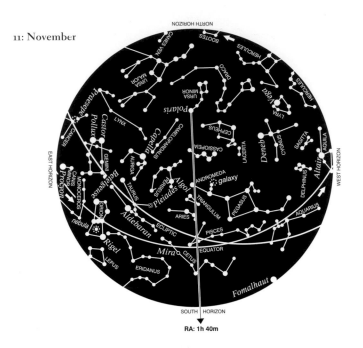

RA: 1h 40m

12: December

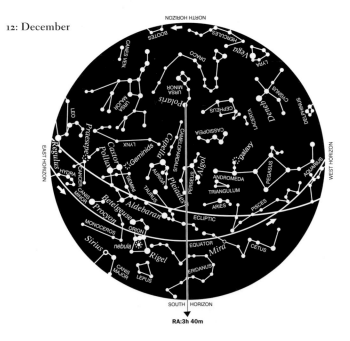

RA: 3h 40m

THE STARS

NUMBER OF NAKED EYE STARS
When we look up at the stars on a clear dark night, their number may seem to be countless. In fact the total number visible to the naked eye at any one time is usually less than 2,000. Faint stars are more common than bright ones and even small binoculars raise the number seen to many thousands. While all the brightest stars are shown on the monthly charts, reproduced on pp.21–26, only those fainter ones needed to make up the constellation patterns can be included on this small scale. These will be 4th magnitude or brighter (see p.35).

DISTORTION ON FLAT CHARTS
The charts show the half of the celestial sphere above the horizon at the stated times. One cannot project a hemisphere on to a circular disc without distortion. The greatest distortion occurs south of the equator where different charts will show some change in the shape of the constellations. This is a compromise and for us at the latitude of British Isles these constellations are less obvious and less important than those farther north, which maintain their basic outlines tolerably well.

ORIGIN OF THE CONSTELLATIONS
Some of the constellation names and figures go back more than 2,000 years. When Europeans started to explore the southern hemisphere new constellations were added containing stars not visible from the north. The inventions of different cartographers were not always convenient. For example there was a large sprawling southern constellation called Argo Navis (the Argonaut's ship), now broken down into smaller constellations but named after parts of the original ship.

Once the main figures made up of naked eye stars, such as Orion, were the constellations and their exact extent was not very important but with the invention of the telescope it became necessary to include ever fainter objects. In 1922 the boundaries of the constellations were redefined by the International Astronomical Union. Every object in the sky now falls into one of these 88 constellations.

NAMES OF THE CONSTELLATIONS AND THEIR MEANINGS
Table 2a shows the names and meanings of the 88 constellations. They vary greatly in extent and some contain few if any bright stars and are

relatively inconspicuous. On the 12 monthly charts from *The Times* there are 49 constellations shown. A further five are excluded because they are small, inconspicuous or there is just not room to fit them in. A further 12 in the south are not shown though parts of them are sometimes visible from the British Isles and the remaining 22 never rise above the southern horizon at our latitude. Table 2a shows into which category each of the 88 constellations falls.

Table 2a	CONSTELLATION NAMES AND MEANINGS

Forty-nine constellations are shown on *The Times* monthly charts:

NAME	MEANING
Andromeda	Andromeda
Aquarius	Water carrier
Aquila	Eagle
Aries	Ram
Auriga	Charioteer
Boötes	Herdsman
Camelopardalis	Giraffe
Cancer	Crab
Canes Venatici	Hunting dogs
Canis Major	Greater dog
Canis Minor	Lesser dog
Capricornus	Goat or sea goat
Cassiopeia	Queen Cassiopeia
Cepheus	King Cepheus
Cetus	Whale or sea monster
Coma Berenices	Berenice's hair
Corona Borealis	Northern Crown
Corvus	Crow
Crater	Cup or chalice
Cygnus	Swan
Delphinus	Dolphin
Draco	Dragon
Eridanus	River Eridanus
Gemini	The twins
Hercules	Hercules
Hydra	Water snake
Lacerta	Lizard
Leo (Major)	Greater Lion
Lepus	Hare
Libra	Scales or balances
Lynx	Lynx
Lyra	Lyre
Monoceros	Unicorn
Ophiuchus	Serpent bearer
Orion	Orion, the hunter
Pegasus	Winged horse
Perseus	Perseus
Pisces	The fishes
Piscis Austrinus	Southern fish
Puppis	Poop of the ship Argo
Sagitta	Arrow
Sagittarius	Archer
Scorpius	Scorpion
Serpens, Caput and Cauda	Serpent, its head and tail
Taurus	Bull
Triangulum	Triangle
Ursa Major	Greater Bear
Ursa Minor	Lesser Bear
Virgo	Virgin

Five small faint constellations not shown on monthly charts:

Equuleus	Little horse
Leo minor	Lesser lion
Scutum	Shield
Sextans	Sextant
Vulpecula	Fox

Twelve southern constellations not shown but parts of which can at times be seen from the British Isles:

Antlia	Air-pump
Caelum	Sculptor's chisel
Centaurus	Centaur
Columba	(Noah's) dove
Corona Austrinus	Southern crown
Fornax	Furnace
Grus	Crane or stork
Lupus	Wolf
Microscopium	Microscope
Pyxis	Compass of the Argo
Sculptor	Sculptor's studio
Vela	Sails of the Argo

Twenty two constellations are never visible from the British Isles:

Apus	Bird of paradise
Ara	The altar
Carina	Keel of the Argo
Chamaeleon	Chamaeleon
Circinus	Pair of compasses
Crux	(Southern) cross
Dorado	Swordfish (or goldfish)
Horologium	Pendulum clock
Hydrus	Water snake
Indus	Indian
Mensa	Table (Mountain)
Musca	The fly
Norma	The square
Octans	Octant
Pavo	Peacock
Phoenix	Firebird
Pictor	Painter, artist
Recticulum	Net
Telescopium	Telescope
Triangulum Australe	Southern triangle
Tucana	Toucan
Volans	Flying fish

Above: *The Milky Way in Scutum, Serpens and Sagittarius, showing clouds of unresolved stars and dark dust lanes.*

Right: *Part of the Milky Way in Cygnus, Cepheus and Lacerta (12 degrees north to south). The hazy disc right centre is an optical effect.*

THE MILKY WAY

Not shown on the monthly charts is the Milky Way, the broad band of light that can be seen stretching from horizon to horizon on a very clear dark night. When the first telescopic astronomers examined the Milky Way, they found much of it to be composed of thousands of faint stars (Figure 2a).

At the time of the November chart it lies from the eastern horizon, up through the zenith and down to the western horizon. It passes through the following constellations: Canis Major, Monoceros, Orion, Gemini, Taurus, Auriga, Perseus, Cassiopeia, Lacerta, Cepheus, Cygnus, Lyra, Sagitta, Vulpecula, Hercules, Aquila, Ophiuchus, Serpens, Scutum, Sagittarius and Scorpius. It continues through the southern constellations Ara, Norma, Circinus, Musca, Lupus, Centaurus, Crux, Carina, Vela, Pyxis and Puppis.

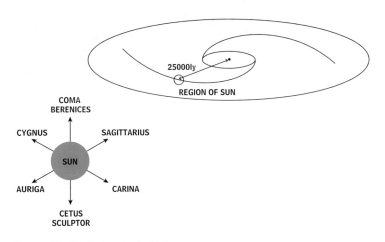

Fig. 2a *The Sun's place in the Galaxy*

THE GALAXY

We now know that our Sun, solar system and all the stars we see with the naked eye lie in a small region of a huge system of stars called the Galaxy. The brightest stars and most dense dust clouds outline the spiral arms of our galaxy. When we look towards the arms we see the band of the Milky Way, but when we look out into space away from the plane of the Galaxy we see darker skies and fewer stars. The Galaxy is up to 100,000 ly across and the centre lies towards the dense star clouds in Sagittarius some 25,000 ly from the Sun. The Galaxy contains perhaps 100 thousand million stars as well as star clusters and bright and dark nebulae (Figure 2a).

*The Great Galaxy in
Andromeda M31, showing it
as it appears in binoculars.*

OTHER GALAXIES AND THE LOCAL GROUP

Galaxies, similar to our own, occur in groups of from a few dozen to thousands. In our Local Group there are about 30 known members of which our Galaxy is one of the largest. The only other galaxy readily visible to the naked eye to northern observers is M31 in Andromeda, also in the Local Group, 2 million ly away which appears as a misty oval of light on a dark moonless night. It is similar if rather larger than our own. It appears about 4th magnitude to the naked eye. In the southern skies are the two smaller irregular-shaped galaxies, the Large and Small Magellanic Clouds which appear to the naked eye like detached parts of the Milky Way. They appear large because they are only 160,000 ly away.

NAMED STARS ON THE TIMES NIGHT SKY CHARTS

There are 19 stars named on *The Times* charts. These are listed in Table 2b in alphabetical order. All but three are among the brightest stars: the exceptions are Polaris (2.0 magnitude) and the two variable stars Algol and Mira. Four other objects are named, two open star clusters Pleiades and Praesepe, the diffuse Orion Nebula and the Andromeda galaxy M31 already described.

THE PLEIADES STAR CLUSTER

The Pleiades, also known as the Seven Sisters because seven stars are generally visible to the naked eye, is an open star cluster in Taurus. To the naked eye it is about a degree across. It is some 400 ly from us and consists altogether of about 100 stars, which studies show are moving apart, confirming the conjecture that the stars formed together as a group at the same time. The stars in the Pleiades, hot and blue in colour, are about 80 million years old. References to the Pleiades in Chinese annals go back more than 4,000 years. The brightest stars are today named after the mythological seven daughters of Atlas.

Table 2b   NAMED STARS SHOWN ON THE MONTHLY CHARTS

NAME	CONSTELLATION	APP.MAG.	ABS. MAG.	SPECTRUM	DISTANCE
Aldebaran	Taurus	0.9	-0.1	K	68
Algol	Perseus	2.1-3.4	-0.2	B	95
Altair	Aquila	0.8	2.2	A	17
Antares	Scorpius	1.0	-4.7	M	326
Arcturus	Boötes	0.0	-0.2	K	36
Betelgeuse	Orion	0.5	-5.6	M	310
Capella	Auriga	0.1	0.3	G	42
Castor	Gemini	1.6	1.2	A	46
Deneb	Cygnus	1.2	-7.5	A	1825
Fomalhaut	Piscis Austrinus	1.2	2.0	A	22
Mira	Cetus	2-10	-	M	94
Polaris	Ursa Minor	2.0	-4.6	F	460
Pollux	Gemini	1.1	0.2	K	36
Procyon	Canis Minor	0.4	2.6	F	11
Regulus	Leo Major	1.3	-0.6	B	85
Rigel	Orion	0.1	-7.1	B	910
Sirius	Canis Major	-1.5	1.4	A	9
Spica	Virgo	1.0	-3.5	B	260
Vega	Lyra	0.0	0.5	A	23

OTHER OBJECTS NAMED:
Galaxy Andromeda (M31) Spiral galaxy of the Local Group
Pleiades Taurus (M45) The 'Seven Sisters' star cluster
Praesepe Cancer (M44) The 'Beehive' star cluster
Nebula Orion (M42 & M43) The Great Nebula in Orion

KEY

APP. MAG. = apparent magnitude, written 'm', how bright the star appears from the Earth.
ABS. MAG. = asbolute magnitude, written 'M', how bright it would appear at 33 light years.
SPECTRUM: A, B stars are bluish-white, F, G yellow-white like the Sun and K, M orange-red in colour.
DISTANCE: in light years at 300,000 km per second.
M = number in Charles Messier's catalogue of nebulous objects.

PRAESEPE STAR CLUSTER
In Cancer there is a larger but fainter star cluster than the Pleiades. Called Praesepe after the Latin for a beehive, its stars are thought to be 660 million years old and at a distance of 530 ly. Its greater age has given time for the stars to disperse more than the Pleiades stars. Praesepe can be seen on a dark night with the naked eye and is a fine sight in binoculars.

Right: *This photograph
(covering about 5 x 3 degrees)
taken on 19 January 1991 of the
Pleiades star cluster in Taurus,
shows many of the stars seen in
general purpose binoculars.
The bright object below centre
is the planet Mars but the
large overexposed image is not
the true planetary disc which is
too small to show on this scale.*

Below: *The open star cluster
Praesepe in Cancer. Covering
10 x 8 degrees this photograph
gives a good idea of the
binocular view.*

THE GREAT NEBULA IN ORION

The Great Nebula in Orion is also visible to the naked eye in Orion's sword below the three stars of Orion's belt. It is 1,500 ly away and is in fact just the brightest part of a huge cloud of dust and gas covering much of the constellation of Orion. This contains such interesting objects as the Horsehead Nebula, a dark cloud projected on to bright clouds beyond. New stars are forming in the Orion nebula which has been the subject of intense study, most recently with the Hubble Space Telescope. Like the Pleiades and Praesepe clusters, the Orion nebula is well seen in the more powerful binoculars and small telescopes.

Main picture: *Orion, printed darkly to show the area as seen with the unaided eye (covering 32 x 22 degrees).*

Inset: *The brightest part of the Orion nebula M42/M43, more as it appears in binoculars or a telescope.*

BRIGHTNESS AND MAGNITUDES

The brightness of stars, planets and other objects is expressed in magnitudes, the lower the number the brighter the object. The Greek astronomer Hipparchus over 2,000 years ago was the first to divide up the naked eye stars into groups of similar brightness calling the brightest 1st magnitude and the faintest 6th magnitude. A 1st magnitude star is 100 times brighter than one of 6th magnitude. Therefore each star is 2.5 times brighter or fainter than the one of the next magnitude. For example a 1st magnitude star is 2.5 times as bright as a 2nd magnitude star, 6.3 times one of 3rd magnitude, 16 times one of 4th, 40 times one of 5th and 100 times one of 6th magnitude. The use of 2.5 rather than 2.0 for each step arises because it is in line with how the eye perceives differences in brightness It should be made clear that magnitude in this context has nothing to do with the size of the star, though some bright stars may happen to be large.

A 6th magnitude star is generally taken to be the faintest that can be seen by a keen eye in a dark sky (though eyes vary and some, especially when young, can see fainter ones). Stars or planets brighter than 1st magnitude are represented by: 0 for 2.5 times brighter than 1st and then –1, –2 as necessary.

Sirius the brightest star is –1.5, Jupiter and Mars can reach –2.8 and Venus –4.6. The full Moon is about –12.5 and the Sun –26.6. This twofold difference in the number between the Moon and Sun is actually a difference of 14 magnitudes: the Sun is brighter than 400,000 full Moons.

APPARENT AND ABSOLUTE MAGNITUDES

The Table 2b gives the apparent magnitudes which is how bright they appear to the eye. The eye cannot detect differences of less than 0.1 magnitude, but instruments can measure the brightness to an accuracy of at least 0.01 magnitude (a hundredth of a magnitude difference).

Since it was discovered that stars differ greatly in brightness, absolute magnitude was devised to compare their real or intrinsic brightness or luminosity by imagining them all to be equally distant. The distance chosen was 10 parsecs or about 33 ly. When the apparent brightness was corrected for distance it was seen that stars varied greatly in intrinsic brightness and that in some cases the ones we see as the brightest appear bright just because they are near.

DISTANCES IN LIGHT YEARS

The approximate distances in light years are also given in Table 2b, and in our small sample of –1.5 to 2nd magnitude stars they range from 9 ly for Sirius to over 1,800 ly for Deneb and in real brightness from 2.6 magnitude for Procyon to –7.5 magnitude for Deneb. This makes Deneb 10 magnitudes brighter than Procyon or put another way, Deneb is as bright as

10,000 Procyons. For comparison the Sun is a G spectrum star of absolute magnitude 4.8, its −26.6 apparent magnitude being due to its nearness, only 8.3 light minutes (149 million km).

SPECTRAL CLASSIFICATION, TEMPERATURE AND COLOUR

Astronomers find out about a star's temperature, composition, rotation and other interesting properties by observing the star's spectrum. Spectral classification is complex but the broad categories are given letters (not quite in alphabetical order) starting with those of highest surface temperature. The principal classes run O, B, A, F, G, K and M. The surface temperatures of these stars in degrees C are in the order of: O (35–45,000), B (15–25,000), A (10,000), F (7,000), G (5,000), K (4,000) and M (3,000). The hottest O and B stars appear bluish, A stars white, F and G are yellow, K orange and M red.

There are no type O stars in Table 2b as they can maintain this high energy output for perhaps only a million years and they are therefore not very common at any one time. The B stars include Rigel which is 50,000 times as luminous as the Sun. While Capella has a G spectrum like that of the Sun, it is a giant star and 100 times brighter than the Sun, which is a G dwarf star. At the other end of the scale are the supergiant red stars like Antares, which, though relatively cool, is still very bright, even at a distance of 330 ly, because of its huge size.

BRIGHTNESS, MASSES AND SIZES OF STARS

Stars range in absolute brightness from more than a million times that of the Sun to only a mere 20 thousandth of the Sun. In terms of mass or weight of stellar material, the range is much smaller, from about a tenth to 50 times that of the Sun. But the variation in size or radius is much greater with the Sun at 1.4 million km and neutron stars only 10 km across while giant stars can reach huge dimensions, for example over 400 million km for Betelgeuse. If placed where our Sun is it would envelop the orbit of Mars. Such supergiant stars have very diffuse atmospheres.

DOUBLE AND MULTIPLE STARS

Not all stars are single but with others form double or multiple star systems. It may happen that two stars appear close but are at greatly differing distances: these optical doubles are not the gravitationally bound systems that interest astronomers. Some pairs can be seen to be orbiting around each other when observed over years through the telescope: these are called visual binaries. Others are too close together to be seen as separate but betray their status through variations in brightness and changes in their spectra. These latter are called spectroscopic binaries. It is from double stars that astronomers find out the masses or amount of matter in a star. Their observation is an important branch of astrophysics.

ALGOL

Algol (beta Persei), also known as the 'Demon Star' from the Arabic 'Al Ghul' (the demon) was studied by John Goodricke in 1782. He established that the fading took place regularly every 69 hours. It fades from 2.1 to 3.4 magnitude every 69 hours taking five hours to fade and another five hours to return to normal brightness. It is an example of an eclipsing binary where two close stars are revolving about each other every 69 hours in an orbit so inclined that one passes in front of the other as seen from the Earth. When the fainter passes in front of the brighter, the light we see is reduced by 1.3 magnitudes. There is a shallower eclipse when the brighter passes in front of the fainter, but this is not detectable to the naked eye. A third star in the system was discovered in 1974 and there may be further discoveries to be made.

MIRA

The other variable star shown on the charts, Mira (omicron Ceti in Cetus) is a long-period pulsating variable. In the case of Mira the period is about 330 days though this is not exactly constant as it is for an eclipsing binary. Mira fades from 2nd magnitude to 10th magnitude but cycles vary both in the magnitudes reached and the interval between maximum and minimum brightness. Because the period averages eleven months, maximum and minimum brightness are on average a month earlier each year, so that maximum brightness can occur for several years when Cetus is in the daytime sky. A late autumn maximum provides a good opportunity to observe Mira brightening and fading again and much of the cycle can be followed using only binoculars. Maximum brightness may be expected to fall in October by the year 2000. Mira varies in brightness by changing its size and surface temperature.

DELTA CEPHEI

This star in Cepheus is nearest the 'C' of Cassiopeia, (see November chart, Fig. 1f (11)). Delta Cephei is another pulsating type variable, though in the case of these stars (Cepheids) the pulsations are of much shorter period and accurately regular. Delta Cephei fades from 3.5 to 4.4 magnitude every 5.4 days. These stars are intrinsically bright and can be recognised at great distances. There is a relation between the absolute brightness and period of variation, so if one can measure the apparent brightness and period, one can calculate the distance. Cepheids have been one of the most important means of finding the distances of the nearer galaxies where they can be recognised on photographs taken with large telescopes. Delta Cephei is not marked on the regular monthly charts as it is a little too faint for easy naked eye observation.

DUST AND GAS CLOUDS, FORMATION OF NEW STARS

Double and multiple stars are common. Stars are formed in clouds of dust and gas like the Great Nebula in Orion and many, perhaps most, start their lives as members of groups. In time, when the dust and gas have been blown away by the radiation from the new stars, we see a star cluster such as the Pleiades or Praesepe. These clusters are among the brightest examples of open or galactic clusters and may be found with binoculars in many parts of the sky but particularly along the Milky Way.

GLOBULAR CLUSTERS

Another form of cluster of which a few are just visible to the naked eye are the globular clusters. In these clusters up to a million stars form a spherical grouping. As with open clusters the stars in a globular cluster have a common origin. Globular clusters are old and are found away from the Milky Way because those that lie in the galactic plane are mostly so far away from us as to be hidden by the dust clouds that lie in the spiral arms of our Galaxy. Globular clusters surround the inner parts of the Galaxy rather like a swarm of gnats.

NOVAE

Nova means new or new star but all novae are actually old stars that have suddenly increased their light output so that a star may brighten ten magnitudes in a matter of hours. A new naked eye star may appear where no star was seen before, but if earlier photographs of the area showing faint stars are examined there will always be found a star in the place now occupied by the nova. The nova is one star in a binary or close double system. A nova readily visible to the naked eye might be seen about every ten years.

On 29 August 1975 the writer discovered Nova Cygni 1975 (unfortunately hundreds of other observers discovered it at about the same time!). It was 2.3 magnitude and almost overhead in Cygnus on a fine summer evening. On going outside it was immediately apparent that the outline of the Swan had changed and having ruled out a passing artificial satellite one knew it must be a nova that had not been visible the previous night. It had been discovered on that same date in Japan where it was dark earlier. It remained visible to the naked eye for a few weeks, fading slowly.

Nova Cygni 1992, showing how quickly a nova fades, (**opposite**) *24 February 1992 (4.3 magnitude) and* (**above**) *10 March 1992 (6.3 magnitude).* Photos: H.B. Ridley

SUPERNOVAE

A supernova is a much more violent outburst caused by the gravitational collapse of a star when energy production in the core can no longer support the outer layers. The brightness may suddenly increase by 16 magnitudes. The single star can emit as much light as 100 million stars like the Sun to become as bright as a whole galaxy. Supernovae are discovered every week or so in other galaxies, because there are so many galaxies, but they are relatively rare in any one, perhaps one per 50 years out of 100 million stars.

The last supernova seen in the Milky Way, our Galaxy, was in 1604, observed by Kepler; this followed one in 1572, observed by Tycho Brahe, which was as bright as Venus and could be seen in daylight.

In 1885 one appeared in the Andromeda galaxy and though 2 million ly away was near naked eye visibility. The most recent naked eye supernova and the first since 1604 was in the Large Magellanic Cloud (LMC) in 1987. This reached 4th magnitude for several weeks but was too far south to be seen from the British Isles.

THE SUN, SEASONS AND TIME

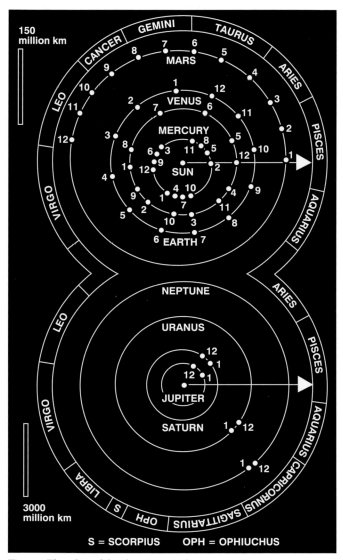

Fig. 3a *The orbits of the planets, 1999 planetary positions*

THE ORBITS OF THE PLANETS

The Figure 3a shows the orbits of the planets from Mercury to Neptune around the Sun. On this scale the Earth's orbit appears circular but it is really slightly elliptical so that at perihelion (January 4) it is 147 million km from the Sun and at aphelion (July 5) 152 million km. The Earth moves slightly more quickly along its orbit near perihelion than at aphelion.

It takes the Earth 365.26 days to complete a revolution about the Sun and return amongst the same stars as seen from the Sun. This period is called the Sidereal Year. The Earth moves forward along its orbit by almost one degree each day as seen from the Sun. The Figure 3a shows that the inner planets revolve about the Sun more quickly than the outer planets. Mercury and Venus, the inner planets, overtake the Earth while the Earth in turn overtakes the outer planets from Mars to Neptune. The effect of this on their visibility is discussed on p.70 on the Planets where Table 6 shows the periods of revolution and other data. Pluto is not shown in Figure 3a as it is never brighter than 13 magnitude and requires a 20 cm or larger telescope to see it.

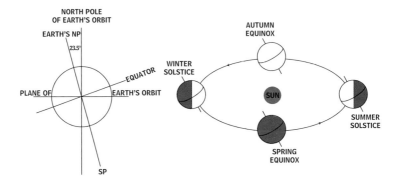

Fig. 3b *The seasons*

INCLINATION OF THE EARTH'S AXIS

The Earth's axis, about which it rotates once every 24 hours, is inclined at 23.5 degrees to the perpendicular to the plane of the Earth's orbit, or put another way, there is a 23.5 degree angle between the north pole of the Earth's orbit and the north pole of the Earth. This is called the Obliquity of the Ecliptic. Because of this, first the northern hemisphere and then, six months later, the southern hemisphere is tilted towards the Sun (Figure 3b).

THE SEASONS

The Sun reaches its most northerly declination of +23.5 degrees on about the 22 June when it is overhead on the Tropic of Cancer, at 23.5

42

degrees N latitude. Six months later it is the southern hemisphere that faces the Sun with the Sun reaching its most southerly declination −23.5 degrees on about 22 December, being overhead on the Tropic of Capricorn at 23.5 degrees S latitude. The tilt of the Earth's axis in combination with its orbital motion round the Sun causes the Earth's seasons.

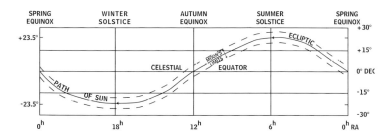

Fig. 3c *The apparent path of the Sun, the ecliptic*

THE SOLSTICES AND EQUINOXES

The date when the Sun is farthest north is the Summer Solstice and that when farthest south the Winter Solstice. The two intermediate points where the Sun crosses the equator are known as the Equinoxes, the Spring or Vernal Equinox being when the Sun crosses the equator from south to north about 21 March, and the Autumn Equinox when it crosses from north to south about 22 September. The dates of all these events vary by a day or so from year to year because there is not an exact number of days in a year and we have introduced Leap Years (Figure 3c).

From our position on the Earth it appears that it is the Sun that moves and not the Earth. The Sun follows the same path against the stars which the Earth follows if it could be seen from the Sun, only six months earlier or later. When we look due south at midnight the Sun is about 180 degrees behind us, below the northern horizon in the part of the sky that will be due south at midnight in six months time. This can be seen from Figure 3a.

Figure 3c shows the apparent path of the Sun against the stars. Although we cannot see the stars in daylight, we know precisely the Sun's position relative to the stars at any time.

Starting at the Vernal Equinox on the equator (RA 00h, Dec 00 deg.) the Sun moves north to the Summer Solstice (RA 06h, Dec +23.5 deg.). It moves into the southern hemisphere at the Autumn Equinox (RA 12h, Dec 00 deg.) reaching its most southerly point at the Winter Solstice (RA 18h, Dec −23.5 deg.), moving north again to the Vernal Equinox and the start of another year.

43

THE ECLIPTIC AND ZODIAC

The path the Sun appears to follow against the star background is called the ecliptic and has been added to the charts in Figure 1f (1–12). The ecliptic is not shown on the regular Night Sky charts to avoid extra clutter as these show the Moon and planets and they are always within a few degrees of it. The Moon is shown there ten times its scale size to show the phases clearly, its centre being its correct position. If drawn to scale the Moon would be about the size of one of the smaller stars. The ecliptic marks the centre of the band of sky within which the Moon and planets are found. This is called the zodiac. The ecliptic passes through 13 zodiacal constellations (see Figure 3a). The constellations from the First Point of Aries are: Pisces, Aries, Taurus, Gemini, Cancer, Leo, Virgo, Libra, Scorpius, Ophiuchus, Sagittarius, Capricornus and Aquarius. Ophiuchus does not appear in the astrologer's zodiac although the planets spend more time in it now than they do in Scorpius. It is still customary to recite the zodiacal constellations starting with Aries (the Ram).

PRECESSION OF THE EQUINOXES

To the naked eye observer the stars can be considered fixed in Right Ascension (RA) and Declination (Dec) but it must be emphasised that the point where the ecliptic crosses the celestial equator moving north, the Vernal Equinox or First Point of Aries, is not fixed but moves slowly westwards by about 50 arcseconds (50") per year. This movement causes slow changes in which constellations make up the zodiac, so that the First Point of Aries is now, after 2,000 years, in Pisces.

Although 50 arcseconds per year is a small amount, in about 26,000 years it adds up to 360 degrees, causing the Vernal Equinox to move right round the equator and back to its starting point. This also causes the north and south celestial poles to move in circles through the northern and southern constellations. Polaris was not always the pole star, we are just fortunate to have a bright star near the celestial pole today. This slow wobbling movement of the Earth's axis, caused by the pull of the Sun and more strongly the Moon on the Earth's equatorial bulge, is called Precession of the Equinoxes (often shortened to Precession) and has to be taken into account when using star catalogues or charts based on say 1950 positions to find a star in the telescope in 1999. But to the naked eye observer the motion is too slow to be a problem in a lifetime.

TIME AND THE SUN

The position of the Sun in the sky still controls our lives broadly speaking and it is convenient to centre our main activity about midday when the Sun is highest in the sky. We saw on p.20 that we can keep accurate time by observing the stars, but if we used them directly after six months midday would become midnight.

The obvious way is to use the Sun directly for time keeping. A sundial does this, when the Sun is visible. But as we saw earlier, the Sun does not move along the ecliptic at a steady speed because the Earth's orbit is not circular. Worse still it does not move eastwards at a steady rate each day because it is moving along the ecliptic and sometimes northwards or southwards at angle to the equator and only at the solstices is it moving exactly parallel to the equator (Figure 3c).

SUNDIAL TIME, LOCAL APPARENT SOLAR TIME
A sundial shows Local Apparent Solar Time but for the reasons mentioned in the previous paragraph, the intervals between successive transits of the Sun, that is the lengths of each day, are not exactly the same throughout the year. This did not matter when people travelled slowly or very little but is useless for accurate timekeeping in the modern world.

When the railways and telegraph came into use there was a need to keep the same time over large areas of a country and the means to distribute time signals by which clocks could be synchronised. The Greenwich Meridian running through the old transit telescope (also known as a meridian circle) at the old Royal Greenwich Observatory in London was adopted as longitude zero (0 degrees) for time-keeping round the world. Accurate clock checks (time signals) are now available from the radio, television and telephone. An inexpensive quartz watch will keep time to a second or two per month.

Mean Solar Time was introduced to provide a uniform time based on the Sun in which each day, hour, minute and second would be the same length as any other. To do this a fictitious, Mean (or average) Sun was introduced that is imagined to travel uniformly along the equator (not the ecliptic). Sometimes it is a few minutes ahead of sundial time and sometimes behind: four times a year the two coincide. When related to the Greenwich meridian this became Greenwich Mean Time. The length of this day is called the Mean Solar Day, and is what is meant when no other type of day is specified. It is also the day of our calendar.

GREENWICH MEAN TIME AND UNIVERSAL TIME
Because the Sun is large and difficult to measure and cannot be seen every day as it crosses the meridian, Greenwich Mean Time (GMT) was actually calculated from observations of the stars, which are easier to time and much more numerous. In astronomy the name Universal Time (UT) is now applied to what was Greenwich Mean Time. They both start at midnight 0h and use the 24 hour clock to 23h 59m 59s after which a new day starts. They are different names for the same time.

Daylight saving schemes like British Summer Time (BST) are in use in many countries. When in force BST is one hour ahead of UT so that 23h BST is 22h UT (or GMT).

Universal Time is calculated from observations of the stars crossing the meridian and is thus directly related to the rotation rate of the Earth. As observations became more accurate it was apparent that the regular motions of the planets were not precisely in tune with our reckoning of time as measured by the Earth's rotation. Ephemeris Time (ET) now known as Terrestrial Dynamical Time (TDT) was introduced to provide a uniform continuous time basis for astrometric observations. Currently it differs from UT by about a minute but it is only important in accurate telescopic observations.

In addition to the long-term slowing of the Earth's rotation, caused mainly by the Moon's tidal pull on the Earth, Universal Time is also subject to small unpredictable variations. Universal Coordinated Time (UTC) was introduced to be run at a uniform rate and be free of all these short-term variations and it is kept to within a second of UT. When the difference approaches 0.9 seconds a Leap Second is added to UTC to bring it closer to UT again. This may be necessary every year or so. UTC is now the time given by radio time signals and generally maintained for civil time distribution. But as the difference between UT and UTC is never as

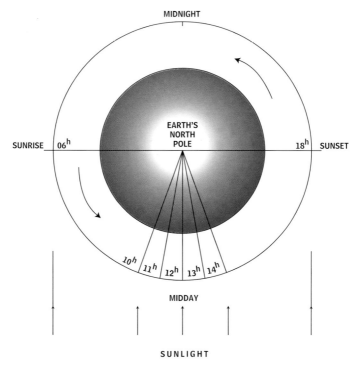

Fig. 3d *Time at different longitudes*

much as a second the two can be treated as equivalent for most purposes. The absolute standard of time keeping is now International Atomic Time (TAI from the French initials). This is the average of several dozen atomic clocks which are accurate to billionths of a second per day. UTC must differ from continuous running TAI because it is tied to UT by leap seconds which in turn is tied to the irregular rotation of the Earth, but the difference between UTC and TAI is maintained so as to be always a whole number of seconds. The difference is currently about half a minute.

TIME AND LONGITUDE
Figure 3d shows for one of the equinoxes how the time by the Sun is earlier in the west and later in the east. The Sun would set at 18h and rise at 06h, the Sun being fixed in direction and the Earth rotating. For example, when it is 12h GMT (noon) in London, the Sun is just rising 90 degrees or six hours farther west in parts of America, where the local time is 06h. Ninety degrees or six hours east of London in the Bay of Bengal the Sun is just setting and the local time is 18h. At times of the year other than the equinoxes the Sun's changing declination and the latitude of the observer complicate calculation of the times of sunrise and sunset but the local times at any one longitude remain the same.

TIME ZONES
The Sun crosses the local meridian 1 hour later for every 15 degrees we move west and 1 hour earlier for every 15 degrees east (Figure 3e). This means that as we travel east we find it is later in the day. Other countries adopt time which suits their needs according to their longitude and large countries like the United States cover several time zones, each differing by a whole hour. As meridians are perpendicular to the equator, the Sun will cross a given meridian as seen from all latitudes at the same moment.

TWILIGHT
Refraction or the bending of light in the Earth's atmosphere makes the Sun, Moon, stars and planets appear slightly higher in the sky than they really are. The effect is only noticeable to the naked eye near the horizon. It amounts to about half a degree at the horizon, the apparent diameter of the Sun or Moon. When the Sun is seen just touching the sea or a low distant horizon it is geometrically just below the horizon. The effect is to slightly increase the duration of daylight (Figure 3f).

A more discernible effect can be the flattening of the Sun as it closes with the horizon. This can cause it to appear oval, more semicircular or jagged in outline. Care must always be taken when looking directly at the Sun but the very red Sun at sunset is usually safe.

If the Earth had no atmosphere it would be dark as soon as the Sun slipped below the horizon but the atmosphere remains lit by the Sun long after the Sun itself has sunk below the horizon, as shown in the Figure 3g.

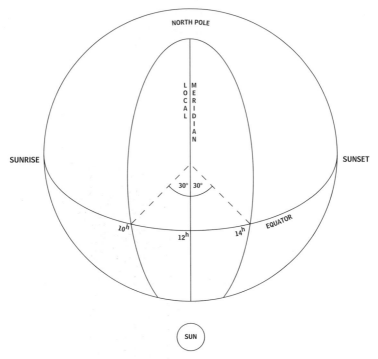

Fig. 3e *Local meridian and time*

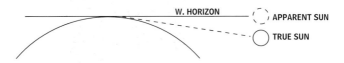

Fig. 3f *Refraction near the horizon*

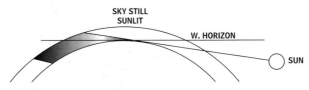

Fig. 3g *Twilight*

DEFINITIONS OF TWILIGHT

Astronomers define three types of twilight depending on how far the centre of the true Sun is below the horizon. How dark it will be at that moment does depend to some extent on how clear the sky is, both in visible clouds and invisible dust, water and ice particles.

When the centre of the true Sun reaches 6 degrees below the observer's horizon Civil Twilight ends. At this time reading a newspaper outside is difficult. When the Sun reaches 12 degrees Nautical Twilight ends: seeing the horizon at sea is difficult. With the Sun 18 degrees below Astronomical Twilight ends and it is as dark at it will get, though it is never completely dark because of airglow, starlight and other causes. The three twilights begin again when the Sun reaches these same depressions at the end of the night (Figure 3g).

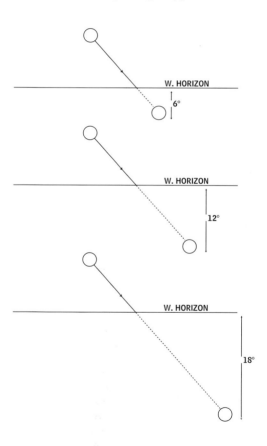

Fig. 3h *Definitions of twilight*

Near the equator twilight is short-lived because the Sun sinks steeply down to the horizon. At the poles twilight can last for weeks as the Sun circles just below the horizon. When darkness comes it lasts for several months. In mid-latitudes the length of twilight depends on the time of the year: it is tabulated for different latitudes throughout the year, for example, in the Astronomical Almanac. The table can be taken as applicable to any year. The times given are for the longitude of Greenwich (o degrees) and the difference in longitude must be take into account in correcting the tabular value for local circumstances. For each degree east of Greenwich sunset, sunrise and twilight occur four minutes earlier and for each degree west four minutes later.

TIMES OF SUNSET AND EFFECT OF LONGITUDE
Even in a small island the time differences are noticeable; for example parts of East Anglia are 1.5 degrees east while Devon, west Wales and central Scotland are 4 degrees west of Greenwich.

Depending on the season, the Sun could rise and set more than 20 minutes earlier in eastern England than in the West Country. At the equator one degree is about 113 km (70 miles), at 50 degrees north or south 72 km (45 miles) and at 60 degrees 56 km (35 miles). At the poles where all lines of longitude converge one can no longer move east or west but at the north pole only south and at the south pole only north and a degree of longitude has shrunk to no distance at all.

As it is not feasible to list times for many different latitudes and longitudes, Greenwich times are usually given in the press and elsewhere but one should be aware that they can vary considerably over the British Isles and much more so over large countries.

CHANGING HOURS OF DARKNESS
Astronomical Twilight lasts all night from about 20 May–20 July at the latitude of London, from 4 May–8 August in Edinburgh and Glasgow and from 22 April–20 August in the Shetland Isles. It gets dark in the evenings and mornings most rapidly in September and October, with the time the Sun is above the horizon being reduced by nearly two hours in each month. The effect on astronomical twilight is even greater with an additional 2.5 hours of real darkness by the end of September.

LENGTH OF DAYLIGHT AT DIFFERENT LATITUDES
The British Isles is spread over more than 10 degrees of latitude from about 50 degrees (Plymouth), London (51.5 degrees), Edinburgh (56 degrees) to the Orkney and Shetland Isles (about 60 degrees). In winter the hours of daylight are fewer in the north than the south, but in summer the reverse is the case. At some point the times of sunrise and sunset must be about equal for all latitudes and this occurs at the equinoxes when the Sun is over the equator.

As examples of how the lengths of the day differ take the four places again. In early January the lengths of time between sunrise and sunset are: Plymouth (8h 10m), London (7h 50m), Edinburgh (7h 20m) and the Northern Isles (6h 05m). So the Sun is above the horizon for two hours longer on the south coast than in the Northern Isles.

In early June these intervals have become 16h 10m, 16h 30m, 17h 15m and 18h 25m, giving Orkney over two hours more sunshine than in the south of England. In practice we are usually more interested in daylight hours than hours the Sun is above the horizon so one has to add in twilight as well. In January the time from the end of Nautical Twilight to its beginning again the next morning is 13h 10m in the south and 14h 10m in the north. This is more than an hour less than the difference in the time the Sun is above the horizon. By early June there are 4h 30m of darkness in the south but none in the north.

EARLIEST SUNSET AND LATEST SUNRISE

The shortest day at our latitudes is on the day of the winter solstice (about 21 December), but the earliest sunset and latest sunrise occur about 10 days before and after the solstice. This arises because we choose to run our clocks at a uniform rate but, as we have seen, the Sun does not move eastwards at a uniform rate.

EQUATION OF TIME

The maximum differences between UT and sundial time occur in November and February. In early November a sundial on the Greenwich meridian would show noon as early as 11h 44m UT while in early February it would show noon as late as 12h 14m UT. The difference is known as The Equation of Time. During December each year sundial noon occurs before 12h UT until the 25 and after 12h UT after that date. The equation takes the form: Mean time = apparent time – equation of time.

On 12 December there is a 13-minute difference between the length of the morning and afternoon, the afternoon starting 6.5 minutes earlier by the Sun than by our clocks. After this date the (slightly) shorter days are offset by the lengthening afternoons and the days start to 'draw out'.

In the morning though the reverse happens as we have shorter days until 21 December and shortening mornings so the Sun continues to rise later until 2 January. Thus the earliest sunset is about the 12–13, the shortest day is the 21–22 December and the latest sunrise not until 2 January.

THE CALENDAR AND LEAP YEARS

The development and complications of the Calendar are beyond the scope of this Companion but it may be confusing that there are so many definitions of the year (and necessarily of the month and the day also). In astronomy they all have their precise uses for example in calculating the position of the Earth, Moon, eclipses and so on. But numerically these values of the year are all within a few tenths of a day of each other and the differences can be ignored by the casual observer.

The Tropical Year (365.2422 mean solar days) is the interval between successive passages of the Sun through the Vernal Equinox, that is successive crossings of the Celestial Equator from south to north. Using this solar year ensures that the seasons do not get far out of line with the calendar. For instance we want to know that each year the time of sunset and the hours of darkness will be the same about the same date. Leap years are necessary because there is an inexact number of days in the solar year. The extra day in a leap year is still added to February (29), the last month in the Roman calendar. In the Julian calendar, introduced by Julius Caesar, all centuries were leap years and in time the calendar began to get out of step with the seasons.

The Pope Gregory's rules, adopted in England in 1752 and in use today, are that years divisible by four are leap years unless they are centuries in which case they are only leap years if divisible by 400. The year 1900 was not a leap year, 1996 was and 2000 will be. Under the Gregorian calendar discrepancies will remain very small for centuries to come.

THE MOON

THE MOON'S ORBIT
The Moon revolves around the Earth in an elliptical orbit at an average distance of 384,000 km but at perigee is at 354,000 km and at apogee 404,000 km. It keeps the same face towards the Earth, turning on its axis once each month. Though we can only see half the Moon's surface at any one time several factors allow us to see a bit farther round first one side and then another. Formations on the Moon near the visible edge or limb sometimes come into view and at other times are hidden so that almost 60 per cent of the Moon's surface can be seen from the Earth at some time or other. The times when each limb is turned more towards the Earth, called libration, are tabulated in astronomical almanacs.

THE MOON'S PHASES
Figure 4a shows how the light from the Sun illuminates half of the Moon's surface. How it appears from the Earth depends on the Moon's position in its orbit round the Earth. New Moon occurs when the illuminated side is turned away from us. As the Moon moves on, anticlockwise as seen from the north pole of the orbit, it passes through the waxing crescent phase reaching half illuminated or first quarter after about 7.5 days. Moving on through the waxing gibbous phase it is full after about 14.75 days when the Moon is on the opposite side of the Earth to the Sun.

The waning gibbous Moon reaches last quarter about 22.25 days after new passing through the waning crescent phase to start again at new Moon after 29.53 days. Because the Moon's orbit is not a circle, the Earth and Moon are moving round the Sun together in an orbit that is not circular and the Moon's orbit is inclined to the Earth's orbit, the intervals between successive phases are not generally exactly a quarter of 29.53 days. This can be seen by checking the published dates and times of the lunar phases. The Moon does not pass in front of the Sun nor the Earth in front of the Moon on every revolution or eclipses would occur every month.

LUNATION, SYNODIC MONTH AND METONIC CYCLE
The average length of this lunation or synodic month from new to new (or other like phase) is 29.53 mean solar days. As there are between 12 and 13 lunations in a year the same phases will not occur on the same days or dates each successive year. An interesting relation is that 235

lunations = 19 years of 365.25 days. This is known as the Metonic Cycle: if a full Moon occurs on a certain date there will be a full Moon on the same date 19 years later. The same relation applies to new Moons and other phases.

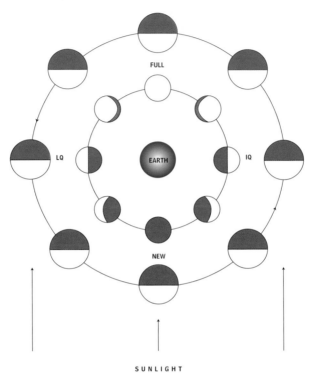

FULL

LQ

EARTH

IQ

NEW

SUNLIGHT

Fig. 4a *The Moon's phases*

SIDEREAL AND SYNODIC MONTH

The Moon orbits the Earth as both orbit the Sun. The Figure 4b shows two positions of the Earth and Moon, A and B. If at A the line Sun-Earth-Moon points to a certain star and the three bodies being in line, the Moon appears at full phase near the star. After the Moon has travelled 360 degrees around the Earth 27.32 days later on average the Moon again appears to be near the same star (The stars are so far away that the directions towards them from both positions of the Earth are essentially parallel). This period of 27.32 days is the Sidereal Month. But the Moon is not yet in line with the Sun and Earth so the phase is not full but about two days before full Moon. The Sun, Earth and Moon come into line after 29.53 days, completing the synodic or lunar month.

Not high resolution photographs, but showing the level of detail that can be seen with ordinary binoculars. Also the eye can accommodate a much wider range of brightness and contrast than film and would see more detail in the brighter parts of the images.

Left: *First quarter Moon on 14 August 1994*
Right: *Gibbous Moon three days before full on 22 April 1994.*

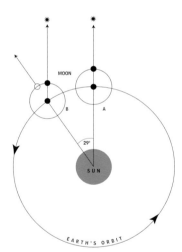

Fig. 4b *The lunar month*

The Anomalistic Month, which averages 27.55 days is the time the Moon takes from perigee to perigee, its closest point to the Earth.

The Moon moves eastwards against the star background by about 13 degrees each day, or by its own diameter in an hour. Thus it comes to the same phase about 29 degrees farther east at each lunation, the full Moon (or any other phase) advancing steadily month by month from the winter constellations to the summer constellations.

MOON'S POSITION AND THE SEASONS

As the full Moon is in the opposite part of the sky from the Sun, that is they are 180 degrees apart, the Moon will appear to be near that part of the ecliptic that the Sun occupied six months previously. When the Sun is high in the northern summer in Taurus and Gemini, the full Moon will be low down in Scorpius and Sagittarius. And in midwinter when the Sun is low down in Scorpius and Sagittarius the full Moon will be high up in Taurus and Gemini.

The monthly charts show the waxing (before full Moon, evening) crescent and gibbous phases, full and the waning (after full Moon, morning) gibbous phases but being timed before midnight but they do not show the waning crescent phases. The charts in *The Times* and *The Times Night Sky* booklet show where the Moon is at each phase in the different seasons. The Moon is said to be in crescent phase between new and first quarter, and again between last quarter and new. Between first quarter and full and between full and last quarter it is gibbous.

In spring the evening crescent is high in the north-west, full Moon near the equator, and the morning crescent low in the south-east. In summer the evening and morning crescents are near the equator with full Moon low in the south, rising in the south-east and setting in the south-west. In autumn the evening crescent hangs low in the south-west while the morning crescent is high in the north-east, full Moon being near the equator. In winter the crescents are again near the equator and the full Moon high in the south, rising in the north-east and setting in the north-west. The Moon follows the ecliptic but whereas it is by definition the exact path of the Sun, the Moon can be a few degrees above or below it.

INCLINATION OF THE MOON'S ORBIT

The Moon's orbit round the Earth is inclined at 5.2 degrees to the Earth's orbit (Figure 4c). The two points where the Moon's orbit passes through the plane of the Earth's orbit, the nodes, move westwards along the orbit completing a circuit of the orbit in 18.6 years.

The most apparent effect of this is in variations in the Moon's altitude and the position of the rising and setting points on the horizon. At one extreme position of the nodes the inclination of the Moon's orbit is added to that of the Earth, 23.5 + 5.2 = 28.7 degrees, while nine years later it has to be deducted, 23.5 − 5.2 = 18.3 degrees to the celestial equator. So the Moon's greatest declination declines over nine years from 28.7 to 18.3 degrees and increases again to 28.7 degrees over the following nine years.

ALTITUDE OF THE SUMMER AND WINTER FULL MOONS

Translated into the altitude above the horizon of the summer and winter full Moons as seen from latitude 52 degrees N, the Moon can cross the

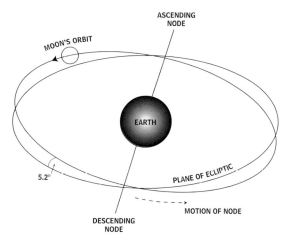

ASCENDING
NODE

MOON'S ORBIT

EARTH

5.2°

PLANE OF ECLIPTIC

MOTION OF NODE

DESCENDING
NODE

Fig. 4c *Inclination of the Moon's orbit*

meridian in winter at an altitude of 38 + 28.7 = 66.7 degrees or 38 + 18.3
= 56.3 degrees. The summer full Moon can be as low as 38 − 28.7 = 9.3
degrees or 38 − 18.3 = 19.7 degrees.

It is well known that the Moon, especially the full Moon, looks
larger when near the horizon. There have been many explanations of
the 'Moon Illusion'. The same effect is seen when a constellation such
as Orion is seen near the horizon and again high in the sky, so it is not
peculiar to the Moon. In fact the Moon is a little farther from us when
near the horizon and so is marginally smaller. Measurements and photographs confirm this is an illusion probably caused by our expectation
that everyday objects seen at low angle are farther away than those overhead,
for example a cloudy sky does not appear as a dome but as a flattened
sheet above us.

RANGE OF THE MOON'S ALTITUDE

The declination of the Moon reached a maximum in October 1988
when it crossed the meridian at an altitude of 67 degrees as seen from
London. The summer full Moon was only 10 degrees above the southern
horizon that year. By 1997 the maximum meridian altitude reached by
the Moon was only 57 degrees but the minimum meridian altitude was
higher at about 20 degrees. The Moon will not be back to its greatest
range in declination until 2006–2007.

At high declination the Moon rises and sets farther round towards
the north in the north-east and north-west in winter and farther round
towards the south in the south-east and south-west in summer. Because
of this it also affects the hours the Moon is above the horizon. Of course

the Moon can reach these declinations and altitudes at phases other than full in these extreme years, though it is the full Moon that is perhaps most noticeable.

TIMES OF THE MOON'S RISING
Since the Moon moves eastwards about 13 degrees in 24 hours it would, if it moved along the celestial equator at a constant speed, rise about 50 minutes later each night. But for reasons we have seen it does not move along the equator nor quite at a constant speed.

EARTHLIGHT
It is easier to see earthlight when the Crescent Moon is high in the sky, on late winter and spring evenings and on autumn mornings. When the Moon is near the new phase to us on the Earth, the Earth is near full as seen from the Moon. Most of the half of the Moon turned towards us is not yet directly illuminated by the Sun. In a clear sky we can see the dark side faintly lit by earthlight. How bright it appears depends partly on the distance of the Moon from the Earth and the cloudiness of the Earth's atmosphere, our clouds reflecting light back to the Moon better than clear skies.

In spring the waxing crescent Moon is seen near the western horizon almost 'on its back' and then earthlight can be strong. This gave rise to the saying about 'the old Moon in the new Moon's arms'.

THE CRESCENT MOON IN FACT AND FICTION
The line of the ecliptic can be visualised by remembering that the light-bulge of the Moon points towards the Sun, here below the horizon. A way of imagining this is to think of the Moon as a bow with an imaginary arrow pointing towards the Sun.

The Moon is often depicted the 'wrong way round' in paintings, advertisements and on Christmas Cards when evening activities are portrayed. Carol singers appear below a crescent Moon bulging towards the left or east, which puts it in the dawn sky! Worse still stars are sometimes shown between the horns of the crescent Moon. Even if most of the Moon's circular disc cannot be seen in the crescent phase it is still there to hide the distant stars beyond as earthlight reveals.

At all latitudes the celestial equator meets the horizon due east and west. At 50 degrees N latitude the celestial equator makes an angle with the horizon at the east and west points of 40 degrees. At the Spring Equinox the part of the ecliptic where the evening crescent Moon is lies north of the celestial equator, making a total angle with the horizon of 63.5 degrees. So the spring crescent new Moon appears high in the west or north-west (Figure 4d).

The situation in the tropics and southern hemisphere is beyond the scope of this Companion, but readers may like to work it out for themselves.

The new Moon (age 36 hours) and Venus on 24 March 1993. In March the new crescent Moon is north of the Sun and sets 'on its back' and earthshine is often strong, giving rise to the old saying about 'the old Moon in the new Moon's arms'. Photo: H.B. Ridley.

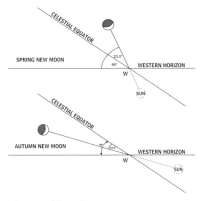

Fig. 4d *Spring and Autumn New Moons*

THE HARVEST AND HUNTER'S MOONS

Another consequence of the slope of the ecliptic near the horizon concerns the rising times of the Moon about full phase in the spring and autumn months. The charts for April and May show the Moon moving south of the equator after full Moon while in September and October it is moving steeply northwards (Figure 4e).

The effect of this is for the Moon after full phase to rise more than 50 minutes later each night in the spring and less than 50 minutes each night in the autumn months. In May the Moon disappears from the evening sky very quickly after full phase. In autumn it can be seen to rise only minutes later each night as it moves farther round towards the north-east. This effect is accentuated when the Moon has a high declination and so is moving south or north at an angle greater than that of the ecliptic. At such times it can be a week before the Moon again rises as much as 50 minutes later each night and the time between successive moonrises can be as little as 24 hours 14 minutes. In the spring, in years when the Moon's maximum declination is large, the time between successive moonrises about the full phase can reach nearly 25 hours and 30 minutes.

The early rising for several consecutive nights of the full Moon in autumn about harvest time gave rise to the name Harvest Moon for the full Moon nearest the date of the Autumn Equinox (about 22 September). It can occur as early as the second week in September or carry over into October, depending on the exact dates of the equinox and of the full Moon. The full Moon following the Harvest Moon is the Hunter's Moon.

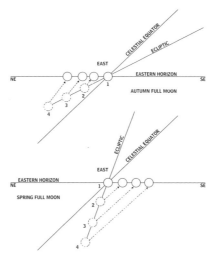

Fig. 4e *Spring and Autumn Full Moons*

MONTHS WITHOUT FULL MOONS

As the average period between one full Moon and the next is close to 29.5 days it can happen that there is no full Moon in February, the only month with less than 30 days. This occurred in 1999 when there were two full Moons in January and two in March but none in February. While it may not happen very often that there is no full Moon in February, it must happen in that month regularly that there is no new Moon or some other phase repeated. But a full Moon or the absence of one seems to be the only phase that arouses comment.

ECLIPSES OF THE SUN AND MOON

GEOMETRY OF SOLAR ECLIPSES

Eclipses of the Sun occur when the Moon comes between the Earth and the Sun. It is a remarkable coincidence that the Moon, only one four-hundredths of the Sun's diameter of 1.4 million km, just covers the Sun which is four hundred times farther away from us.

SOLAR ECLIPSES

As the Moon orbits the Earth it usually passes above or below the Earth-Sun line but every so often it comes directly between us, and some part of the Earth sees a solar eclipse. Figure 5a shows the general situation. It is not drawn to scale but shows that the Moon's dark or umbral shadow covers only a small part of the Earth.

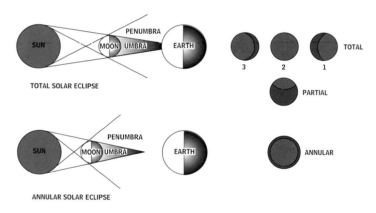

Fig. 5a *Total, annular and partial solar eclipses*

At mid-eclipse the Sun, Moon and Earth lie in a straight line so all solar eclipses take place at new Moon. But the Moon is moving eastwards as seen from the Earth's north pole, so the shadow it casts on the surface of the Earth also moves rapidly eastwards. Because the shadow is projected on to a sphere its path, which may be several thousand kilometres long, will not generally be a straight line on a map. The path from which a total or annular eclipse can be seen is usually quite narrow, often less than 100 km.

WARNING ABOUT OBSERVING THE SUN

It is dangerous to look directly at the Sun with the naked eye and very dangerous with any optical instrument including a camera viewfinder. During totality it is safe but precautions must be taken during the partial phases as even a thin crescent of sunlight may cause eye damage. See p. 101.

The Sun totally eclipsed on 24 October 1995. Photographed from India by R.J. McKim.

TOTAL SOLAR ECLIPSES

At any point along the path of totality observers will see first a partial eclipse with the Moon's eastern limb (edge of the Moon's disc) encroaching on the Sun's disc from the west, so a northern observer will see a notch appear in the right hand side of the Sun. The total phase can last from a few seconds to a maximum of about 7.5 minutes and then there is a partial phase again as the Moon moves towards the eastern edge of the Sun's disc. Within a broad band, usually a few thousand kilometres wide, north or south of the path of totality a partial eclipse will be seen.

People travel all over the world to witness a few minutes of totality. What makes a total eclipse special is that with the Sun totally obscured by the Moon, it becomes much darker and the brighter stars and planets can be seen near the Sun. The temperature drops sharply and wildlife reacts to these changes. As the Moon covers, and a few minutes later, uncovers the brilliant white 'surface' of the Sun (the photosphere) it can be seen shining through valleys on the Moon's limb like a string of starlike points along the limb: this effect is known as Baily's Beads. Often as the limb of the Sun becomes visible again, this flashes out at one point giving rise to the Diamond Ring effect. Both phenomena are better seen at the end of totality as the eye is then used to the lower level of light and is better prepared for when it is about to happen. Both effects last only a few seconds, as the Moon's shadow races eastwards.

During the seconds or minutes of totality, it may be possible to see chromosphere and solar prominences. The chromosphere is a very thin layer of gas just above the photosphere (6,000 degrees C) but at a higher temperature: it is transparent in white light. Prominences look like red flames rising from the Sun's surface. While not flames they are at a high temperature of about 10,000 degrees C constrained into loops and columns by magnetic fields, often associated with sunspots. They can be seen with very narrow band filters at any time outside an eclipse.

The Corona is hotter, up to 2 million degrees, white in colour and much fainter. Without very special and expensive instruments, usually located on mountain tops, it can only be seen from the Earth's surface during a total eclipse. The ghostly white light can be seen to extend out into space several solar radii and may appear in long streamers.

ANNULAR SOLAR ECLIPSES

If an eclipse occurs when the Moon is close to its farthest from the Earth, at apogee, it may be too small to cover the Sun completely, especially when the Earth is near perihelion and the Sun appears at its largest. Then the Moon's shadow does not quite reach to the Earth and an observer on the central line of the eclipse will see the black disc of the Moon surrounded by a bright ring or 'annulus' of the Sun's surface. Such an eclipse is called an annular solar eclipse. The sky around the Sun will be too bright to enable the corona and other phenomena to be seen. Again observers on either side of the track will see a partial eclipse.

REQUIREMENTS FOR A TOTAL OR ANNULAR ECLIPSE

We saw earlier that the Sun's distance from the Earth varies slightly and so does the Moon's distance from the Earth. This causes the apparent sizes of the Sun and Moon to vary also.

The Sun's apparent diameter varies between 31.6 and 32.6 arcminutes while the Moon's apparent diameter varies between 29.4 and 33.6 arcminutes. If the Moon appears smaller than 31.6 minutes a solar eclipse cannot be total and is either partial or annular: if the Moon's diameter is greater than 32.6 minutes, a solar eclipse cannot be annular, it is either total or partial. It can occasionally happen that the apparent diameters of the two bodies are so close that the eclipse will be total for part of the eclipse track but annular for other parts. This arises because the distances from different observers along the eclipse path to the Moon are not all equal due to the curvature of the Earth's surface. In these circumstances totality where seen will be brief.

PARTIAL SOLAR ECLIPSES

Partial solar eclipses are more common than total ones because they only require that there will be a partial eclipse seen from somewhere on the surface of the Earth: it will not be total or annular anywhere. The Moon is seen to hide part of the Sun's face to an extent that may be anything from a small notch in the limb to near-totality.

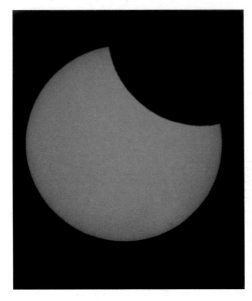

Partial solar eclipse on 12 October 1996, photographed through gaps in the cloud sheet, showing the Moon obscuring the north-west portion of the Sun.

THE SAROS ECLIPSE CYCLE

For a solar eclipse to occur the Moon must be crossing the ecliptic, the path the Sun follows round the sky. This crossing point is one of two nodes of the Moon's orbit. It takes the Sun 346.62 days between successive passages through the same lunar node: this is known as the synodic period of the node of the Moon's orbit. Nineteen synodic periods total 6,585.8 days while 223 lunations from new Moon to new Moon equal 6,585.3 days. The two intervals are nearly equal and the interval of 6,585 days (18 years and 11 days) is known as the Saros.

The importance of the Saros is that if there is to be an eclipse of the Sun, both Sun and Moon must be near the same node: 6,585 days later they will again both be near the same node, 19 synodic periods having elapsed and it will be new Moon again as 223 lunations have elapsed also. Thus with conditions the same a solar eclipse will again occur, though it need not be of the same kind, total, annular or partial, and it will not usually be visible from the same locations. The same is true of lunar eclipses; after 18 years and 11 days (or 10 days if there are five leap years) eclipses will be repeated.

Total eclipse of the Moon 18–19 January 1954.
A series of exposures taken approximately
every 10 minutes before and after totality.

The full Moon at mid-eclipse, when the whole
Moon was immersed in the umbra, was barely
visible to the naked eye.

NUMBER OF ECLIPSES IN A YEAR
The least number of eclipses possible in a year is 2, both solar. The greatest number is 7 of which either 4 or 5 are solar and either 3 or 2 are lunar.

LUNAR ECLIPSES
In a lunar eclipse, the three bodies have to be more or less in line as for a solar eclipse but this time the Earth is between the Sun and the Moon, so all lunar eclipses occur at full Moon. Because the strip of the ecliptic where eclipses can occur is some days long, solar and lunar eclipses often occur two weeks apart, that is at new Moon and full Moon of the same, or successive lunations.

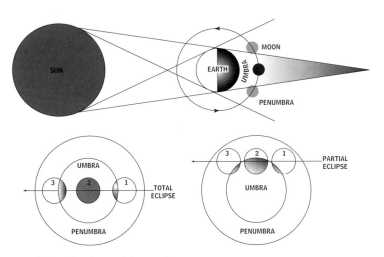

Fig. 5b *Total and partial lunar eclipses*

LENGTH OF TOTALITY
Figure 5b shows the general situation at a lunar eclipse. The Moon is moving eastwards into the Earth's shadow and so the dark or umbral shadow of the Earth appears to move from east to west over the Moon's face. The shadow is larger than the Moon so a lunar eclipse takes longer than a solar eclipse. The longest duration will be if the Moon passes centrally through the Earth's shadow when totality may last more than 1.5 hours. If the Moon passes near the edge of the Earth's shadow, totality will be shorter.

PARTIAL LUNAR ECLIPSES
A partial lunar eclipse occurs when part of the Moon passes outside the Earth's dark umbral shadow. The remaining part lying only in the penumbra will appear quite bright, though not as bright as the

uneclipsed full Moon. Penumbral eclipses usually go unnoticed as the Moon passes only through the Earth's penumbral shadow and is only slightly darkened.

OBSERVABILITY OF LUNAR ECLIPSES

Unlike a solar eclipse, where every place sees a slightly different eclipse at different times, with a lunar eclipse we all see the same eclipse, with times of totality the same. The altitude of the Moon above the horizon will vary from place to place, and half the Earth will be turned away from the Moon and will not see it. The whole event may last several hours.

The appearance of the Moon during an eclipse varies depending on the state of the Earth's atmosphere. The atmosphere bends and scatters some sunlight so that some light reaches the Moon even at mid-eclipse. Generally the light is reddened so that the eclipsed part of the Moon appears orange, red or copper coloured but may be grey. Sometimes the Moon is quite bright at mid-eclipse but at other eclipses may be almost invisible to the naked eye. In a telescope or binoculars one can usually see the maria or 'seas' and often craters on the Moon as well.

DEFINITION OF AN OCCULTATION

In astronomy, when one body comes between the observer and another body there is said to be an occultation. An eclipse of the Sun should more properly be called an occultation of the Sun by the Moon but it has long been known as a solar eclipse. And there is a shadow of the Moon cast on the Earth by the Sun which would be visible from the Moon, so any selenites present would see an eclipse of the Earth.

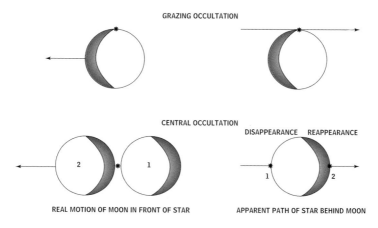

Fig. 5c *Lunar occultations*

OCCULTATIONS OF STARS AND PLANETS BY THE MOON

The meaning of an occultation was explained above. As the Moon moves round the sky it must pass in front of many stars. In fact the number of stars brighter than say 6th magnitude that it covers each month is measured only in a score or two. Occasionally it passes in front of a 1st magnitude star or more rarely a planet. As the path of the Moon is continuously changing it can hide a star in several consecutive months and then not do so again for several years. (Figure 5c).

BRIGHT STARS THAT CAN BE OCCULTED BY THE MOON

Only the following 1st magnitude stars lie within the path of the Moon at some time in the lunar cycle and can be occulted: Aldebaran, Regulus, Spica and Antares. The planets can all be occultated at some time or other.

OCCULTATIONS OF A PLANET BY ANOTHER PLANET

Occultations can also occur when a planet passes in front of a star. These events are even rarer, even with quite faint stars because the planets are so small compared with the Moon, and they move much more slowly against the sky. Occultations of planets by other planets are very rare indeed, but occultations and eclipses of satellites by their parent planets occur regularly and those involving the four great satellites of Jupiter can be watched with quite a small telescope.

VISIBILITY OF OCCULTATIONS

As with solar eclipses, the track on the Earth's surface from which an occultation by the Moon can be observed is not very wide, so a star or planet that is occulted from say Edinburgh will not necessarily be occulted from Plymouth. Because of their much greater distances from us, occultations of stars by planets are usually visible over a wider area.

THE PLANETS

Table 6	PRINCIPAL DATA ON THE PLANETS				
THE TERRESTRIAL PLANETS	Mercury	Venus	Earth	(Moon)	Mars
Diameter (km)	4878	12104	12756	3476	6794
Mass (Earth = 1)	0.055	0.815	1.000	0.012	0.107
Sideral rotation period	58.65d	243d	23h 56m	27.32d	24.62h
Mean distance (AU) (million km)	0.387 57.91	0.723 108.90	1.000 149.60	— —	1.524 227.94
Sidereal period	88d	224.7d	365.26d	27.3d	687d
Synodic period	115.88d	583.92d	—	—	779.9d
Eccentricity	0.206	0.007	0.017	0.05	0.093
Inclination of orbit	7.00	3.39	0.00	5.15	1.85
Number of satellites	0	0	1	0	2

THE GIANT PLANETS	Jupiter	Saturn	Uranus	Neptune
Diameter (km)	142,800	120,000	52,000	48,400
Mass (Earth = 1)	317.8	95.3	14.5	17.2
Sideral rotation period	9.84h	10.23h	17.8h	16.1h
Mean distance (AU) (million km)	5.20 778.3	9.54 1427	19.19 2870	30.06 4497
Sidereal period (yr)	11.86	29.46	84.01	164.8
Synodic period (day)	398.9	378.1	369.7	367.5
Eccentricity	0.05	0.06	0.05	0.01
Inclination of orbit	1.30	2.48	0.77	1.77
Number of satellites	16	18	18	8

POLAR ORBIT DIAGRAM OF THE SOLAR SYSTEM

The polar orbit diagram of the solar system appears on the inside cover of the annual Night Sky booklet and shows for the year the positions of the planets in their orbits as seen from the north pole of the solar system (Figure 3a, p.41).

There are two diagrams because there is such a large difference in the size of the orbits of the terrestrial planets and the giant planets that it is impossible to show them clearly on a single chart of manageable size. The principal data on the planets is given in Table 6, including the mean distances of the planets from the Sun in Astronomical Units (AU) and kilometres. An AU is the mean distance from the Earth to the Sun or 149,597,870 km.

For the terrestrial planets the positions are plotted for the first of each month, but for the giant planets only for the first of January and December because they move relatively slowly.

One can see from the upper diagram that the planets nearest the Sun move round their orbits most quickly. This, combined with the fact that the circumference of their orbits is smaller the nearer the Sun they are, means that they complete more orbits in less time. The inner planets overtake the Earth and the Earth overtakes the outer planets.

EARTH'S LONGITUDE
The position of the Earth along its orbit is defined as its Longitude (this has no connection with the use of longitude on the Earth's surface). As the Earth's orbit is taken as the standard plane its latitude is always zero. But for the other planets their positions are given in longitude and latitude north or south of the Earth's orbital plane (the ecliptic).

In the lower diagram the Earth's orbit is so small that we can assume that the positions in the sky of the giant planets as seen from the Earth are the same as if seen from the Sun. We can draw a pair of lines from the Sun to a planet to see in which constellations it will appear during the year. If the names of the constellations appear to run backwards in the lower part of the diagram to how we see them on the monthly charts, remember that we are looking at them from the outside instead of from the inside of the circle.

DEPENDENCE ON THE EARTH'S POSITION
In the top diagram the Earth's orbit is larger than that of Mercury and Venus but smaller than that of Mars. The directions of these three planets depend very much on where the Earth is at a particular time.

FIRST POINT OF ARIES
First of all though, there is a direct link shown between where the Sun appears to be as seen from the Earth and the background constellations. On 21 March, the Spring Equinox, the Sun appears to lie in the direction of the large arrow, towards the First Point of Aries which is the intersection of the ecliptic and the celestial equator already explained on p.43. This direction was towards Aries 2,000 years ago but has now moved westwards into Pisces due to the Precession of the Equinoxes. The Sun is in the daytime sky and the constellations Leo and Virgo will be in the southern night sky at midnight, when we on the Earth have our backs to the Sun.

USE OF THE POLAR DIAGRAM
Six months later at the Autumn Equinox the Earth will lie on the arrow and Pisces will be in the southern sky at midnight. From this we can see that we cannot observe Jupiter in March and April because it will be in the

same direction as the Sun, near conjunction, but in September they will be in opposite directions and Jupiter will be at opposition, crossing the southern meridian at midnight and visible for much of the night. From this diagram one can see for the year where the planets will be and whether they will be observable.

MORNING AND EVENING STARS

Planets which cross the local meridian before midnight are called evening stars while those that cross after midnight are morning stars. It will be clear that for the outer planets they will change from being morning stars to evening stars as they pass through opposition. With the inner planets they remain either morning or evening stars throughout the whole apparition.

APPARENT MOTION OF THE OUTER PLANETS

We can see that the apparent motion of the outer planets, in which we include Mars with the four giant planets, will be quite different from that of the inner two planets, Mercury and Venus.

If we were observing from the Sun we should see the planets moving steadily eastwards or anticlockwise as seen from the north pole. Planets with almost circular orbits would move at an almost constant rate, while Mars with its more elliptical orbit would move noticeably faster when near perihelion and slower when at aphelion.

DIRECT AND RETROGRADE MOTION

As seen from the Earth the motion of all the planets is still eastwards taken over a whole year but as the Earth is moving too, and more quickly than any of the outer planets, it will overtake them around the time of opposition. Then for a time the planets will appear to move backwards against the stars as in Figure 6a. We say they have changed direct motion for retrograde motion or just that the planets are retrograding. The points at which the motion changes to retrograde or changes back again to direct motion are called stationary points and a planet at one is said to be stationary (Figure 6a).

If the orbits of the planets lay exactly in the same plane we should see the planet move back exactly along its old track between the stars, but because of the slight inclination of the orbits, the path is usually above or below the old path. In some cases it forms a very flattened 'S' shape and sometimes a flat oval loop.

INFERIOR AND SUPERIOR CONJUNCTIONS

Returning to the polar diagram in Figure 3a, the path across the sky followed by the two inner planets Mercury and Venus is more complicated than that for the outer planets. Mercury and Venus can be in conjunction with the Sun in two positions, when passing between the Earth and Sun called inferior conjunction and when on the far side of the Sun, called superior conjunction. For example if Venus were at 12 when the Earth was at 2 it would be at inferior conjunction but if Venus were at 1 it would be near superior conjunction (see also Figure 6b).

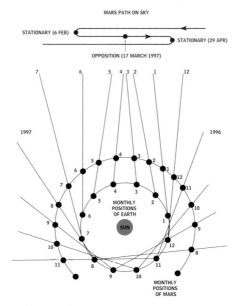

Fig. 6a *Direct and retrograde motion near opposition*

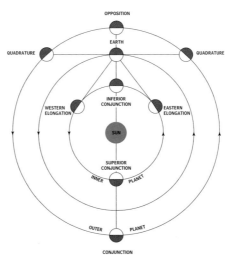

Fig. 6b *Inner and outer planets: terms explained*

GREATEST ELONGATION

Mercury and Venus can never be at opposition however as, being inside the Earth's orbit, they can never be in the opposite part of the sky to the Sun. The farthest either planet can appear to be from the Sun is called greatest elongation, west in the morning sky and east in the evening sky. Mercury's greatest elongations vary from 18–28 degrees because of its more inclined and elliptical orbit while Venus reaches 46–47 degrees in its almost circular orbit. Venus can be nearly twice as far from the Sun in the sky as Mercury and so, being in a darker sky and much brighter too, it is much easier to see.

APPARENT PATH OF AN INNER PLANET

In Figure 6c we see the path on the sky round the midday Sun of an inferior planet such as Mercury or Venus. The diagram shows eastern and western elongation and superior and inferior conjunction points. Either planet may pass directly behind the Sun but then it will not be observable so close to the bright disc. However, if it passes directly in front of the Sun, a transit will occur.

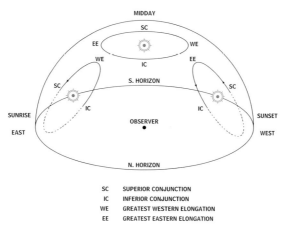

SC	SUPERIOR CONJUNCTION
IC	INFERIOR CONJUNCTION
WE	GREATEST WESTERN ELONGATION
EE	GREATEST EASTERN ELONGATION

Fig. 6c *Inner planets: morning and evening apparitions*

TRANSITS OF MERCURY AND VENUS

During a transit the planet will be seen as a black disc, very small in the case of Mercury. The planet appears not unlike a sunspot, moving slowly across the Sun's face from east to west, taking up to several hours to do so. Transits take place at the nodes of the orbit and are not frequent, the planet usually passing above or below the Sun as seen from the Earth. Transits of Mercury take place in May or November with transits in May 1970 and with the next in May 2003. November transits

Top left: 7 October 1991, Venus (R) and Jupiter (L) with Regulus (above) forming a right angled triangle. Top right: 28 December 1992. The Moon with Venus below and Saturn below-right in the evening sky. Bottom left: 5 August 1995. The gibbous Moon with Jupiter below.

took place in 1973, 1986 and 1993 with the next on 15 November 1999 and 8 November 2008. Not all transits are visible from all places where the Sun is above the horizon and in this respect they are like solar eclipses. How they appear depends on whether Mercury crosses the Sun centrally or near to the north or south limb.

Transits of Venus are even more rare, occurring either singly or in pairs 8 years apart. The last were in 1874 and 1882 and the next will be in 2004 and 2012 after which we shall have to wait until the year 2117.

WARNING ABOUT OBSERVING NEAR THE SUN

Observing the Sun without proper precautions can be dangerous and lead to blindness. Before attempting to watch a transit, solar eclipse, sunspots or any event close to the Sun refer to p.101.

AN APPARITION OF AN INNER PLANET

Returning to Figure 6c, starting on the right-hand side of the diagram, from superior conjunction the planet emerges from twilight into the evening sky, and being on the far side of the Sun it appears to move slowly. It slows again around greatest elongation as it is coming towards the Earth. It then moves more quickly into evening twilight to pass between the Earth and Sun at inferior conjunction. It is then in morning twilight (left-hand side of the diagram) to reappear in the morning

Venus passing Jupiter in the evening sky. **Top left**: 18 February, **Top right**: 20 February, **Bottom left**: 22 February, **Bottom right**: 24 February 1999. *Venus below Jupiter on 18 and 20 and above on the 22 and 24 February. The night of closest approach was cloudy!*

sky, now west of the Sun, slowing down as it moves away from us towards greatest western elongation. It then begins to approach the Sun again to be lost in morning twilight at superior conjunction. Note that the inner planets may be at conjunction to the north or the south of the Sun, and the path is not strictly elliptical as seen from the moving Earth. It is more flattened than shown in the figure, the planet passing nearer the Sun. It takes Mercury 116 days to complete this cycle and Venus 584 days.

CONJUNCTIONS OR CLOSE GROUPINGS OF STARS AND PLANETS

The term conjunction is used for planets coming in line with the Sun (or more correctly having the same longitude). It is also used to describe two or more planets having the same longitude or being close in the sky. There are relationships as to how often this can happen, having regard to the differing times taken for the planets to overtake each other. Suffice it to say that the close grouping of two or more bright planets or a planet with the Moon, perhaps also with a bright star, is a very attractive spectacle, though bearing in mind their greatly different distances, of no real astronomical importance. Notice of conjunctions is given in the annual Night Sky booklet and monthly Night Sky notes.

FAVOURABLE APPARITIONS OF INNER PLANETS

Not all apparitions are equally favourable for observers. Generally the planets are better seen in the tropics where they can pass directly overhead and rise and set steeply to the horizon. Some apparitions are better for northern observers and others for those in the southern hemisphere. There are also longer term effects which favour one hemisphere over the other. These and the general visibility of each planet are of sufficient interest to make it best to consider each in turn.

MERCURY

Mercury is never visible well clear of the horizon in a completely dark sky from the latitudes of the British Isles but it can be as bright as the brightest star Sirius (–1.5 magnitude) and is quite easily seen with the naked eye for a week or two during the most favourable apparitions. These occur between February and April in the evening and September and November in the morning. In midsummer the all-night twilight makes visibility difficult so from mid-May to mid-August the sky is very bright near the horizon. Midwinter apparitions can be quite good if Mercury's orbit carries it north of the Sun.

Because Mercury's orbit is markedly elliptical the planet is further from the Sun at some elongations than at others (from 18–28 degrees) and this with the rapidly varying brightness makes the favourable opportunities both rather infrequent and brief, perhaps lasting only two weeks two or three times each year.

The sidereal period of Mercury is 88 days after which it has completed an orbit relative to the stars, but the synodic period, when it overtakes the Earth and the new cycle of conjunctions and elongations begins again, is 116 days. So in a year there are just over three cycles, or a total of six apparitions, three in the morning and three in the evening. However as three cycles total only 348 days and there are 365.24 days in a year, the cycles do not repeat on the same dates each year, some must overlap the year-ends and the circumstances for observers of the planet differ from year to year.

When Mercury is visible in the evening sky it is east of the Sun and in the area the Sun will move into in a few weeks time. So in late winter and spring when the Sun is moving north, Mercury will be north of the Sun. Thus an evening apparition at this time of the year can be favourable, but how favourable depends on other factors too. As mentioned above,

Mercury's orbit is much more elliptical than that of Venus and its apparent distance from the Sun or greatest elongation is more affected by where it is in its orbit in relation to the position of the Earth. Mercury's orbit is also more steeply inclined to that of the Earth than is that of Venus, adding another variable.

Morning apparitions occur with Mercury west of the Sun, so that the planet occupies the part of the sky where the Sun was a few weeks earlier. In late autumn and early winter the Sun is moving south and so Mercury will be lagging behind it or be further north than the Sun.

Mercury is a rough, cratered, airless body like the Moon and reflects most light towards us about full phase, that is near superior conjunction. But then it is near the Sun and at its farthest from us. As it moves towards greatest elongation it changes towards half phase (like the first quarter Moon) and as it comes closer before passing between the Earth and Sun it has its mostly unilluminated face towards us and we see only a crescent.

Mercury varies in apparent size by a factor of about three between superior and inferior conjunction, reflecting its changing distance from the Earth. But this is more than offset by the change in phase. The magnitude varies from −1.5 or even −2.0 near superior conjunction to +3 or fainter when approaching inferior conjunction.

So Mercury will be brightest at the start of an evening apparition and at the end of a morning apparition and faintest towards the start of a morning apparition and towards the end of an evening apparition. However the altitude of the planet above the horizon and darkness of the sky have a major effect on the visibility of the inner planets. The planet's brightness is important but it may be more readily visible when not at its very brightest but in a darker sky. One needs an unobstructed skyline, preferably away from bright lights and industrial haze. Observing from high ground helps here. A clear frosty or stormy sunset may be better than a calm, fine misty one especially if the planet is very low.

VENUS

Venus is the most conspicuous object in the night sky after the Moon and about five times brighter than Jupiter at its brightest. It is much brighter than Mercury and varies in brightness comparatively little with changing distance and phase. About the same size as the Earth, its cloud covered atmosphere reflects much of the sunlight falling on it. It is easily visible to the naked eye with the Sun above the horizon and it can cast a shadow on a dark night.

Much of what has been said about the visibility of Mercury applies to Venus, though Venus is simpler in that its orbit is almost circular and is almost in the same plane as that of the Earth.

The average or mean distance of Venus from the Sun is 108 million km while that of the Earth is 150 million km. This means that when it is on the far side of the Sun, Venus is 108 + 150 = 258 million km from us while at its closest it is only just over 40 million km. At inferior conjunction it comes closer to us than any other major planet. This sixfold variation in the distance means a 6 times difference in the apparent size of the planet. When closest it appears about 1 minute of arc across (one sixtieth of a degree) or one-thirtieth the diameter of the Moon. The variation in brightness is only about a magnitude from −3.8 to −4.6, just over twofold. Greatest brilliancy occurs near greatest elongation, unlike Mercury, the difference being caused by a combination of the greater change in apparent size and the all-enveloping atmosphere that scatters light in all directions.

Venus takes 225 days to complete a revolution of its orbit (sidereal period) compared with 88 days for Mercury. Venus goes through the same cycles as Mercury but the synodic period is five times as long, 584 days against 116 days for Mercury. The apparitions are much longer, mainly for this reason, though the visibility is further increased by the planet's greater brightness enabling us to see it low down and in bright twilight.

The result of taking 584 days (about 19 months) from one superior conjunction to the next is that two cycles take just over 3 years. If it took exactly 18 months every fourth year would begin a repeat performance. The circumstances of every third apparition will be similar, though not quite the same, as after 3 years we have to wait another 2 months for Venus to come to superior conjunction and the times of inferior conjunction, greatest elongations and the general visibility of the planet will also be two months later.

Recalling what we said about Mercury being north or south of the Sun and the time of the year when this occurred, if Venus reaches greatest eastern elongation during the first few months of the year, it will lie north of the Sun, in the part of the sky where the Sun will be up to two months later. Thus it will set farther round to the north than the Sun and be seen at a higher altitude in a darker sky than if it were south of the Sun.

When Venus is west of the Sun, in the morning sky, it is in the part of the sky where the Sun was up to two months before. If it comes to greatest western elongation during the autumn it will be north of the Sun and rise farther round towards the north.

The year 1996 was a very good one for seeing Venus from the British Isles and it was possible to see it with the naked eye almost continuously throughout the year, in the evening sky until the end of May and from early July until the end of the year in the morning sky. With superior conjunction having taken place in August 1995 the circumstances could hardly have been better.

A superior conjunction in August–October leads to a good apparition the following year. Venus was east of the Sun in the evening sky and ahead of the Sun as it moved north from January to June. With inferior

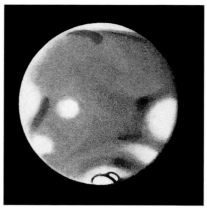

The Moon and Venus seen in daylight at 14.45 UT on 31 December 1997. Drawn to scale, notice the phase of Venus was clearly discernible using only 9 x 35 binoculars. Venus was 16 days before inferior conjunction and 47 million km from the Earth while the Moon was only 374,000 km away. Both show a thin crescent phase. Mars (not shown) was 2 degrees to the east.

The planet Mars on 21 April 1982, showing the north polar cap at the bottom. Drawn by R.J. McKim using a 320 mm aperture refractor.

conjunction in June, the planet was again north of the Sun as it followed it south during the second half of the year. So we had a good evening apparition followed by a good morning one all in one year.

In 1995 when the previous superior conjunction was in January 1994 Venus was close to the Sun and unobservable from the British Isles for much of the year, with only a short showing in January–February and December. The year 1997 was also rather poor with superior conjunction in early April and Venus remaining low and in evening twilight until November–December, nor was 1998 favourable for northern observers. The year 1999 has been good though not quite as favourable as 1996 with superior conjunction having taken place rather late on 30 October 1998.

MARS

Mars is about half the size of the Earth and moves in a more elliptical orbit that averages 228 million km from the Sun but can be as little as 208 million km at perihelion or as much as 248 million km at aphelion. Its closest to the Earth around opposition can vary from 56.3 million km at perihelic oppositions to 100 million km at aphelic oppositions. Mars appears almost twice as large at perihelic oppositions which occur in August or September. Unfortunately for northern hemisphere observers Mars will then be south of the equator and lower in the sky. Winter oppositions put the planet high in Taurus or Gemini but the planet is then farther from us and appears smaller.

Mars takes 687 days (about 23 months) to complete a revolution round the Sun, during which time the Earth has completed 1.9 revolutions. The next opposition will occur when the Earth overtakes Mars: this takes another 3 months making Mars' synodic period 780 days. So oppositions of Mars occur at average intervals of 2 years 2 months. The most favourable oppositions occur at intervals of about 15 years, the next August opposition being in 2003.

The brightness of Mars at opposition can vary from about –1 to –2.8, equal to Jupiter. When far from the Earth it fades to around +1.7 magnitude. Mars can be identified by its orange-red colour.

The 1996–1997 apparition can be taken as an example of an unfavourable opposition (Figure 6a). Mars was at conjunction with the Sun in March 1996 becoming visible in Taurus in July at 1.5 magnitude, passing down into Virgo at 0.5 magnitude by December. It reached a stationary point in Virgo on 6 February 1997 retrograding through Virgo to be at opposition on the 17 March at –1.0 magnitude. After a stationary point on 29 April its motion was again eastwards or direct moving down through Scorpius by October at 1 magnitude. It remained low in the west after sunset from August 1997 until conjunction on 12 May 1998. So in broad terms one sees Mars in the night sky every other year. For a year or more it is rather close to the Sun in the sky and not easily observable. The periods when Mars is close enough for detailed study really only last a few weeks every two years.

JUPITER
Jupiter has a sidereal period of 11.9 years but a synodic period of 399 days. This means that Jupiter moves right round the ecliptic passing through all the 13 zodiacal constellations in 12 years, or through 30 degrees each year. It takes the Earth an extra month to catch up with Jupiter so oppositions take place about a month later each year and about 2 hours of Right Ascension farther east along the ecliptic.

Being 770 million km from the Sun the difference in brightness between opposition and conjunction varies less than in the case of Mars, from about –2.8 to –1.8 magnitude.

The four great or Galilean satellites of Jupiter are easily visible in the more powerful binoculars and their motion about the planet can be followed in detail in a small telescope. They range from 4.6 to 5.6 in magnitude but the glare of the planet makes them more difficult to see than isolated stars of the same brightness. Io, the innermost of the four, takes only 1.8 days to orbit the planet, so its motion is easily detected in a few minutes.

The four satellites vary in diameter from 3,138 km for Europa to 5,262 km for Ganymede, the largest satellite in the solar system.

The planet Jupiter on 11 January 1990 showing the great red spot (upper right). Drawn by R.J. McKim using a 220 mm aperture reflector (south at top).

The planet Saturn on 21 April 1982 showing the north face of the rings. Drawn by R.J. McKim using a 320 mm refractor (south at top).

SATURN

Saturn's sidereal period is 29.5 years so it takes this long to come back among the same stars. The synodic period is only 370 days. Oppositions take place about 2 weeks later each year or about 12 degrees (48 minutes of RA) farther east each year. In 1996 Saturn crossed the equator into the northern hemisphere, where it will remain until 2010.

Saturn averages 1,427 million km from the Sun. The brightness varies slightly because of this change in distance. It is also affected by the angle at which the huge ring system is inclined towards us and the Sun. The rings, seen edge on, contribute little or no light. The southern face of the rings will be facing us until the planet moves south again.

The largest satellite, Titan, is visible in the smallest telescopes. The major feature of the planet is the ring system. Twice in every orbit or about every 15 years, the plane of the rings passes through the Sun which means that first the north and then the south side is illuminated. For a few days the rings are edge on to the Sun and may disappear from view even to observers with large telescopes. About the same time the Earth will also pass through the ring-plane. Depending on the position of the Earth in its orbit, this may happen only once or three times.

The Earth moved from north to south of the ring-plane on 22 May 1995 and on 11 August the Earth moved from south to north. On 19 November the rings were edge on to the Sun and on 12 February 1996 the Earth finally moved from north to south again and the rings began to open out, fully illuminated. The rings may be invisible even in large telescopes about these times.

The next single ring-plane crossings will be in 2009 and 2025 and we must wait until 2038–39 for the next triple crossing.

URANUS

Uranus has a sidereal period of 84 years and a synodic period of 370 days. Its motion along the ecliptic is very slow. The brightness varies slightly reaching about 5.6 at opposition, bright enough to see with the naked eye, though identifying it against the stars is very difficult. It has a mean distance of 2,870 million km.

NEPTUNE

Neptune moves even more slowly taking 165 years to circle the sky, its sidereal period. The synodic period is only 367.5 days. It is 7.9 magnitude and varies little with changing distance. It has a mean distance of 4,497 million km.

PLUTO

Pluto is not shown on the monthly charts as it is never brighter than 13 magnitude. It is a small icy body only 2,300 km in diameter and in a highly elliptical orbit inclined to the Earth's orbit by 17 degrees. Its sidereal period is 248 years and its mean distance from the Sun is 5,913 million km, though this varies from 4,500 million km to 7,500 million km, so at perihelion (1989) it can be just inside the orbit of Neptune. From 1979 to 1999 Neptune has been the most distant of the planets. Pluto will be at aphelion in the year 2113. It has one satellite, Charon, which has a diameter of about 1,200 km.

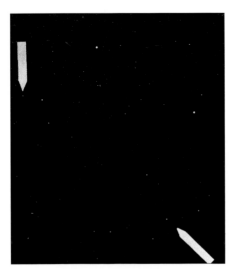

Neptune (upper left) 7.9 magnitude and Uranus (lower right) 5.7 magnitude on 2 August 1992. While the two planets were only 3 degrees apart on the sky, Neptune was 4,380 million km away while Uranus was 'only' 2,790 million km distant. The two planets are easy to see in binoculars but much harder to identify as they appear starlike, their tiny discs being too small to discern without a telescope.

COMETS, ASTEROIDS AND METEORS

COMETS

Most comets are faint objects visible only on photographs or with electronic imaging devices now widely used by professional and amateur astronomers. Each year up to half a dozen comets come within the reach of small telescopes and binoculars. Every few years a comet becomes easily visible to the naked eye.

Bright naked eye comets are rather infrequent and in the last 50 years the Southern comet of 1947, Eclipse comet of 1948, Arend-Roland and Mrkos of 1957, Ikeya-Seki of 1965, Bennett of 1970, West of 1976, Halley of 1986, Hyakutake of 1996 and Hale-Bopp of 1997 spring to mind. Not all were well seen from our latitude however. Most bright comets are to be seen in evening and morning twilight and remain very bright for only a week or two. Interference by moonlight and twilight reduce the opportunity to see them further and one can see little from heavily light-polluted areas. The spring of 1997 brought comet Hale-Bopp into view as a bright naked eye object, brighter than 1st magnitude, and easily visible in twilight and from town centres. It remained a conspicuous object, above the horizon all night, for several weeks, and there must be few who did not see it. It then moved into the southern hemisphere where it was again widely observed.

Top left: *Comet Hale-Bopp photographed from north-east Essex on 28 March 1997 showing* (left) *bluish plasma or ion tail and* (right) *the yellowish dust tail.*
Top right: *Comet Hale-Bopp on 6 April 1997. This higher resolution photograph shows detailed structure in the plasma tail.*

THE NUCLEUS

The heart of a comet is the solid nucleus, a frozen mass of ice and dust. Halley's comet had its nucleus closely observed by the Giotto spacecraft which found it to be about 8 by 15 km in size. Some comets may be much larger; Hale-Bopp probably about 40 km and Hyakutake perhaps less than 4 km across.

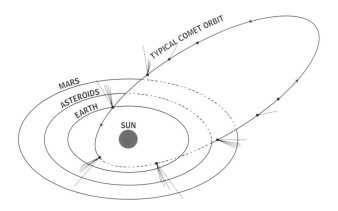

Fig. 7a *Development of a comet as it nears the sun*

DEVELOPMENT OF THE HEAD AND TAIL

As a comet approaches the Sun the ice begins to melt or sublimate releasing gas and dust particles which form the head or coma and may move on to form a visible tail which points away from the Sun. The coma may be several hundred thousand km across and the tail tens of millions of km long. But the total amount of matter in the visible coma and tail is minute by planetary standards; it is very thinly spread (Figure 7a).

BRIGHTNESS

The brightness of a comet is expressed in magnitudes but while a close comparison can be made of one star with another, it is much more difficult to compare a fuzzy comet with a point-like star. The convention is to take only the head and ignore the tail as much as possible. Even so with a comet like Hyakutake, with a head over a degree across, comparison, even with the naked eye, is difficult. One can use spectacles or bifocals, or even binoculars looked through from the wrong end, to try to bring the star and comet nearer to the same size. But comet magnitudes are not as accurate as star magnitudes and generally the larger the comet appears, the more difficult it is to estimate its brightness. Differences of a magnitude or more are common even among experienced comet observers.

THE TAIL

The tail points away from the Sun but while the narrow gas, ion or plasma tail seen well in comet Hyakutake, is usually straight and relatively narrow, a dust tail can be strongly curved and very wide like an open fan, as for example in comet West in 1976 and comet Hale-Bopp in 1997. Perspective and our viewpoint play a major part in how we see a comet as with such a huge object, the various parts are at greatly differing distances from us.

ORBITS AND NUMBERS OF COMETS

The orbits of comets vary in shape and size. Unlike the planets they do not all move round the Sun in the same direction, and some, like Halley's comet, have retrograde orbits. There are in the order of 200 known comets with periods of revolution around the Sun of less than 200 years. These are called periodic or shortperiod comets and move in elliptical orbits. Most other comets have very large orbits which can barely be distinguished from parabolas though some are very long ellipses and others slightly hyperbolic. These mostly have periods measured in thousands of years and many have orbits so large it is meaningless to assign periods in years to them. The number of known comets with well defined orbits is increasing every year as new more sensitive and computerised instruments are being used to search for them. Many are being found as a result of searches for asteroids with Earth-crossing orbits.

The asteroid Eros (7.5 magnitude) passing kappa Geminorum (3.7 magnitude) on 24 January 1975. The movement of the minor planet during the 40 minute exposure is shown by the length of the trail.
Photo: H.B. Ridley

ASTEROIDS OR MINOR PLANETS

The essential aspect of a comet is that it undergoes some degree of out-gassing near the Sun. Those few asteroids, also called minor planets, that have been closely examined by spacecraft or studied from the Earth are rocky bodies that show no sign of a coma or atmosphere whether temporary or permanent. Sometimes it is difficult to know to which category a newly discovered small body belongs. Some comets may eventually become indistinguishable from asteroids. If this is so then some asteroids may be comets which no longer release dust and gas as they approach the Sun.

NUMBERS OF ASTEROIDS

Thousands of asteroids now have well determined orbits and several thousand have now been named. Only Vesta is bright enough to be seen readily with the naked eye and as it appears like a star, tracking it down is not easy. The asteroids cannot really be considered naked eye objects. Most are small and fainter than 15th magnitude. Thousands of other asteroids are known but not yet sufficiently well observed to be named and thousands more wait to be discovered. The main asteroid belt is between Mars and Jupiter but some orbits pass inside that of the Earth.

EARTH‒CROSSING ORBITS

Attention has been focused recently on asteroids with Earth-crossing orbits. Many more have been discovered, some by accident and others by deliberate searching. Apollo and Aten type asteroids can pass within the Earth's orbit and pose a threat of collision. While a close approach is interesting, an impact could be very serious indeed with widespread damage, tidal waves and possible climatic changes. The chances of a collision are small but it has happened in the past and will one day happen again. In 1996 another asteroid passed at just beyond the Moon's distance. Very faint objects only a few hundred metres across are now being discovered; these are necessarily near the Earth when discovered and move rapidly across the sky.

METEORS

On any moonless night of the year one may see shooting stars or meteors. At certain times of the year they are much more numerous. A meteor is the visible effect of a small particle entering the Earth's atmosphere at up to 70 km per second. The object heats the surrounding thin air, which glows, often leaving a train persisting for several seconds. It becomes visible at a height of around 90 km.

METEOROIDS AND METEORITES

Before entering the atmosphere these particles, called meteoroids when in space, are in orbit round the Sun. Most originate in comets and are released as the Sun heats up the nucleus. Others may be the pulverised dust from collisions between asteroids. There are also larger iron and rocky stones that sometimes reach the ground as meteorites.

METEOR STREAMS

Particles released from comets continue to orbit the Sun in similar orbits to the comet, giving rise to streams of meteoroids or meteor streams as they have come to be called. If the orbit of a stream intersects the Earth's orbit we may see a meteor shower. Meteor showers recur about the same dates each year as the Earth reaches the same point in its orbit about that date. If the meteors are concentrated in one part of the orbit, as often happens with a newly formed stream, there may be few meteors seen until the Earth again encounters the denser part of the stream. The gravitational influence of the planets distorts the meteor stream orbits so that streams that miss the Earth now may one day meet it, giving rise to a new shower, while existing streams may be perturbed into new orbits so that they may disappear temporarily or forever.

SPORADIC METEORS

Apart from the shower meteors there are those that do not appear to belong to any recognised stream; these are called sporadic meteors and may be seen, usually in small numbers, on any clear night.

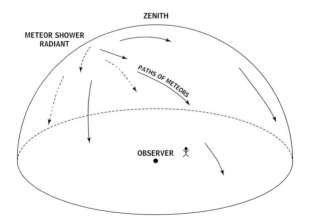

Fig. 7b *Apparent paths of meteors from the radiant*

VISIBILITY OF METEORS

Generally the presence of the Moon, if between first and last quarter, is very detrimental to the number of meteors one can see. A broken or partly clouded sky is preferable to an overall thin cloud or haze for seeing all but the brightest meteors.

The meteors that meet the Earth are travelling in parallel paths; the way in which they spread out over the sky like the ribs of an umbrella is an effect of perspective (Figure 7b). Meteors may be seen in any part of the sky, those near the radiant giving shorter trails than those far from it. The apparent speed depends partly on the real speed of entry and partly on our viewpoint. Brightness varies with speed and size and composition of particle and so varies from meteor to meteor and stream to stream. Some meteors leave trains like lines of sparks. Generally the meteor is visible for up two seconds only, often much less.

Table 7	PRINCIPAL METEOR STREAMS				
MONTH	SHOWER	LIMITS	MAXIMUM	ZHR	COMMENTS
January	Quadrantids	2-4	3	10-50	Blue and yellow
April	Lyrids	19-25	21	10	Good 1982
August	Perseids	11-14	12	60	Some trains
October	Orionids	20-26	21	20	Fast, trains
November	Taurids	End Oct-Nov	3	10	Slow, bright
November	Leonids	15-20	17	0-100	Very fast
December	Geminids	12-14	13	60	Medium speed

NOTES:

1. The limits over the dates on which meteors from the shower can be seen are longer than given above: they may be noticeable to the casual observer between the dates shown.
2. The date of maximum activity varies by a day or two from year to year.
3. The zenithal hourly rate is higher than the number that can be expected when the radiant is lower in the sky.
4. The positions of the radiants are shown on the appropriate monthly chart: see Figure 1f (1-12).

MAJOR METEOR SHOWERS

The Table 7 lists the more important meteor showers visible from the British Isles. Shower meteors appear to radiate from a small area of sky called the radiant. A stream is usually called after the constellation in which the radiant lies (Geminids, Perseids), smaller streams by a nearby star in a constellation (Eta Aquarids), and occasionally by the name of the parent comet (Giacobinids after comet Giacobini-Zinner).

Showers such as the Perseids and Geminids give good displays every year, a single observer seeing 60 per hour on a good night. Some showers like the Leonids give much stronger showers at times when the denser part of the stream is near the Earth. In the case of the Leonids this happens at 33 year intervals, though not all displays are equally strong. Very strong displays or meteor storms usually last minutes rather than hours and are easily missed by being at the wrong longitude when they occur. Then it may be daylight, the radiant low or the Moon obtrusive.

DAYLIGHT SHOWERS
Meteors leave ionised trails in the atmosphere which reflect radio waves so daylight showers can now be studied. Many new showers have been discovered since the early application of wartime radar equipment in the late 1940s.

HOURLY RATES
The numbers of meteors seen depends on the numbers entering the atmosphere at any time but also on the altitude above the horizon of the radiant. When this is very low we cannot expect to see many meteors. Assuming a constant rate, more will be seen as it rises higher. In meteor studies corrections are made for this by calculating a zenithal hourly rate (ZHR). It is this figure usually quoted in tables of meteor showers but it can be taken as an expected upper limit; the average observer will see fewer unless the real rate is higher than expected.

A Leonid meteor, showing an end burst of about -3 magnitude, passing below the bright stars Castor and Pollux on 17 November 1998.

FIREBALLS AND BOLIDES

Fainter meteors are usually more common than brighter ones but, as one cannot be looking exactly at the point where the meteor will appear except by chance, more brighter ones may be seen. For fast meteors the eye rarely picks up those fainter than 3rd or 4th magnitude. Meteors may be −1, −2 or brighter. The bright ones are less common and can be spectacular. Sometimes, as in the 1998 Leonids display, the meteors seen for several hours were predominantly bright fireballs.

Meteors brighter than Venus, say −5 magnitude, are called fireballs or sometimes bolides. The brightest may have bright trains of sparks or leave a glowing column of air along their path. They may cast shadows at night and be easily visible by day when a column of dust may be left along the path.

If the body were to fall to Earth, the iron or stone recovered would be a meteorite. Meteorite falls are sometimes accompanied by booms or noises like thunder but in any one area they are rare, though over the whole Earth some fall every day, mostly into the sea.

If accurate observations can be made from two or more points a few tens of kilometres apart, an accurate path can be calculated for the fireball or meteor, which, with the velocity, can enable an orbit to be calculated. In the case of a fireball, it may lead to a meteorite being recovered. The velocity is obtained from photographs through a rapidly rotating shutter.

OTHER NIGHT SKY PHENOMENA

THE AURORA

The Aurora Borealis (northern lights) occurs within the Earth's atmosphere at heights of from 80–550 km when particles from disturbances on the Sun are guided towards the Earth's magnetic poles and interact with atoms high in the atmosphere. Ultra violet and other radiation reaches us at the speed of light, in just over eight minutes but particles take much longer, 30 hours or more typically.

It is beyond the scope of this Guide to describe the aurora in detail or the reasons for it. It is often associated with large sunspots seen near the centre of the Sun's disc, solar flares, unusual radio reception or changes in the magnetic field. It may recur on a second night and sometimes mention is made in the News, weather forecasts or daily newspapers of unusual conditions or sightings of the aurora in the south. In 1995–96 the Sun was near sunspot minimum with no detectable sunspots for weeks at a time. The next sunspot maximum is due about the year 2000 so strong displays of the aurora may become more common again.

Displays are most common in bands centred on the Earth's magnetic poles, for example through Canada, Greenland and the Arctic, but at times of unusual activity these move southwards towards the equator. The Aurora Australis is the corresponding phenomenon seen in the southern hemisphere.

AURORAE IN THE BRITISH ISLES

In the British Isles the aurora borealis is quite common in Shetland and Orkney and much of Scotland and readers from the north will no doubt have seen it many times. But it may be seen also at least as far south as the English Channel at times of strong activity. These remarks are addressed to those who may not yet have seen it.

COLOUR AND APPEARANCE

While in the far north the predominant colour seen is green, in the south displays are usually white or white and red. Great displays more often take place at times of maximum solar activity but may occur at any time. If you have a reasonably dark low horizon it is worth having a look each clear night for any unusual brightness along or above the northern horizon. Bear in mind that from mid-May to late July there will be a twilight glow along the northern horizon, even at midnight.

A TYPICAL AURORAL DISPLAY FROM SOUTHERN BRITAIN

Displays appear and disappear with great rapidity. A typical sequence is for a low arch of white light to be seen above the northern horizon, often from the north-northwest to the north-northeast. The arch may rise higher in the sky with dark sky beneath it and rays of white light extend up towards the zenith. The rays are sometimes red and large shapeless patches of red or white may be present. Changes take place in seconds and eventually the display retreats towards the horizon becoming a pale glow or arch. It is worth waiting to see if the display will be repeated as quite often it starts all over again the same night, perhaps several times.

THE AURORAL DISPLAY OF MARCH 1989

Although moonlight can spoil a display, the light is often so bright as to be quite evident even with a bright Moon. For example, on the evening of 13 March 1989 there was a bright auroral display seen from many parts of the British Isles where skies were clear. It was in progress as it got dark but the writer, in north-east Essex, had cloudy skies until 22h but from then the display could be followed until cloud moved in again at 23h 30m.

The Moon was just past first quarter and quite high in the south-west. At times the northern sky was so bright that it seemed as if dawn were breaking with trees seen black against the sky. Reddish rays and greenish arcs came and went sometimes reaching the zenith. Several times activity died away to the northern horizon, only to start all over again. All but the brightest stars were lost in the bright sky, which made the waxing Moon seem quite pale. It was less colourful than the one seen here on 28 October 1961 but the best may have been missed because of cloud and in 1961 the Moon had not risen.

A bright aurora is a magnificent sight even from the south of the British Isles and one not to be missed as really good displays with clear weather may happen less than half a dozen times in a lifetime.

THE ZODIACAL LIGHT AND GEGENSCHEIN

A few decades ago all astronomy books included mention of the zodiacal light and gegenschein. Because of the ever increasing brightness of the night sky it is now difficult to see them. A very dark site is needed, preferable on high ground or looking over the sea. The zodiacal light is a faint glow in the form of a semi-ellipse lying along the ecliptic seen after the end of twilight in the evening or before the beginning of twilight in the morning. It is caused by reflected sunlight from interplanetary dust particles.

It is easiest to see when the ecliptic makes a steep angle with the horizon in the evening in the spring and in the morning in the autumn. In this respect the requirements are similar to those for the new and old crescent Moons but it cannot not be seen if there is any moonlight. Running from the evening zodiacal light to the morning zodiacal light is an even fainter zodiacal band. Opposite the Sun in the sky and therefore south at midnight is the gegenschein, a circular area of faint light, too faint to be seen against the Milky Way. These phenomena are only worth searching for in unusually clear conditions from very dark sites with dark-adapted eyes. Once a common sight they have become a curiosity in many parts of the world due to increasing light pollution.

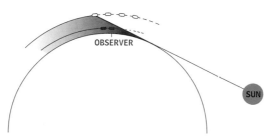

Fig. 8a *Noctilucent clouds*

Top left and right: *Noctilucent clouds on 2 June 1992 at 22.00 and 22.30 UT. Seen from north-east Essex looking north, they show how near the horizon the clouds can appear from latitude 52 degrees north and the changes that can occur in 30 minutes.*
Left: *Noctilucent clouds on 27 June 1966 at 21.45 UT seen from north-east Essex looking north-west.*

NOCTILUCENT CLOUDS

Noctilucent clouds are clouds in our atmosphere that occur at heights of about 80 km. The highest ordinary clouds are usually below 16 km. Because of their great height they are illuminated by the Sun long after even the higher of our ordinary clouds have been lost in darkness (Figure 8a).

In the British Isles the conditions for seeing them are best for a few weeks around the summer solstice when the Sun is less than 18 degrees below the northern horizon and twilight lasts all night even in the south.

It is worth having a look each clear night from late May until late July if the northern horizon is unobstructed. As the Sun sinks lower and moves below the horizon round towards the north, the ordinary clouds will appear black against the twilight sky. If any noctilucent clouds are present they will appear bright against the sky, often bluish-white or silvery-white in colour. They form rows of tiny cloudlets, a herringbone pattern, or parallel rows of clouds. From southern England they will be only a few degrees above the horizon, moving slowly round towards the north-east as the night progresses.

There is some reason to suppose that they are becoming more common and both 1994 and 1995 for example saw displays visible from all parts of the British Isles. A completely dark site is not needed as the clouds are seen in summer twilight: however the eyes should be shielded from nearby bright lights if possible.

On 2 December 1957 at 05.50 UT, the USSR Sputnik 2 carrying the dog Laika passed over south-east Essex.

ARTIFICIAL EARTH SATELLITES

On any clear night the sky from many parts of the world is criss-crossed with the lights of aircraft. There are also other moving lights caused by the reflection of sunlight on artificial Earth satellites. These are in orbit at upwards of 200 km and the lowest orbit the Earth in 90 minutes. Those in higher orbits take longer. There are now thousands of objects in orbit round the Earth, many too faint to be seen with the naked eye.

VISIBILITY FROM THE BRITISH ISLES

From the British Isles only those satellites in orbits inclined to the equator by 50 degrees or more can pass directly overhead. Those in lower inclinations are hidden by the curve of the Earth unless they are in high orbits. Satellites in polar or near polar orbits will also pass over us. Satellites, other than those in polar orbits, are launched eastwards so will rise in the western sky and set in the eastern sky. Polar satellites will pass from north to south or south to north. Some satellites have orbits between these extremes (Figure 8b).

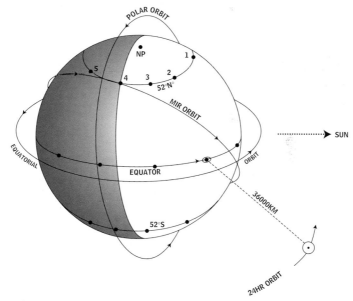

Fig. 8b *Artificial Earth satellite orbits*

REQUIREMENTS FOR VISIBILITY

An artificial satellite is only visible if it is illuminated by the Sun. But if the sky is too bright it will not be observable from the ground. Therefore it is in twilight that most satellites are seen when it is dark enough on the ground for us to see them and the brighter stars, but the Sun has not sunk too far to cut off sunlight from the satellite. From the satellite the Sun would still be above the horizon, the curved edge of the Earth. A satellite in the evening sky will generally pass into the Earth's shadow before it reaches the eastern horizon and a satellite in morning twilight will emerge from shadow into sunlight as it moves eastwards. Except in midsummer when the Sun is not far below the northern horizon, satellites are not visible right across the sky or at every pass.

MIR SPACE STATION
The largest and brightest satellite in orbit in 1999 is the Russian Mir space station. Launched in 1986 its life has been extended several times. As the inclination of the orbit is 52 degrees it can pass overhead in southern England. Although it must cross the British Isles every day this will not usually happen when the conditions are right for observing it. It may be midday or midnight. Visibility will depend on the position of the orbit in relation to the twilight zones and position of Mir in its orbit. Often it will be visible on two consecutive passes about 90 minutes apart, for example passing from west to east and 90 minutes later from west to south, or perhaps once from south-west through our zenith to south-east. Only part of this track may be in sunlight, but as Mir can be as bright as Jupiter, –2 or brighter, it is easily seen. It is due to be decommissioned in the year 2000 and efforts will concentrate on making the parts that will not fully burn up fall into the sea.

INTERNATIONAL SPACE STATION
Up to 1998 the Space Shuttle undertook several rendezvous missions to Mir to gain experience with the building of the International Space Station. The first two units of the ISS were delivered into orbit during 1998 by the Shuttle and a Russian unmanned launcher. Many loads will be delivered and assembled during the next few years. The ISS will be visible from the British Isles and similar northern latitudes and may be expected to become much brighter than Mir as additional units are added. The Shuttle is visible to the naked eye too and will no doubt be seen near the ISS from time to time.

A single bright light crossing the twilight sky may not be too obtrusive but we must hope that the threat of other more sinister schemes for space mirrors to light up whole towns and orbiting signs advertising popular products will somehow be averted. If not, much serious astronomical observation at present carried out from the Earth's surface may soon become impossible.

OBSERVING THE NIGHT SKY

Aircraft condensation trails photographed over north-east Essex on an otherwise predominantly clear day. Now observations of the Sun are seriously impaired by this form of atmospheric pollution, as night-time observations are by lights.

CLEAR SKIES

Observing is a skill that benefits from practice. An experienced observer knows what to look for and how best to see it. Even with the naked eye we can learn to see more by becoming familiar with the night sky and the ever changing conditions in which we have to observe it.

Most night-time objects are best seen in a clear dark sky. The clarity depends not just on the weather but on the presence or absence of industrial haze, wind-borne dust and sand (which can sometimes reach the British Isles from the Sahara) and volcanic dust high in the atmosphere after major eruptions of which there have been several in recent years. Industrial haze may be an almost permanent feature in some areas while volcanic dust can remain aloft for several years and has been the cause of months of colourful sunsets around the world.

Aircraft condensation trails are an increasing problem to astronomers. They are much more noticeable by day but are often present at night being more easily seen when there is a Moon. Living away from airports is not always the answer as in this respect low-flying aircraft are not the culprits. Contrails form at greater heights and a few narrow trails can soon expand to cause a large part of the sky to become covered with amorphous thin cloud making it useless for observing. As light pollution has ruined the night sky, contrails are now ruining the daytime sky.

The weather varies from completely cloudy when no observing is possible, to an absence of cloud. Even when there is no obvious cloud the sky may not be really clear. Astronomers distinguish between clarity or transparency and steadiness of the air or seeing. The best mountaintop sites have many nights a year when both are excellent but nearer sea level in the British Isles we are lucky to get a night when one is good while to have both good on the same night is comparatively rare.

Good transparency allows us to see faint objects. Good seeing allows us to see fine detail. Therefore good seeing is much more important to observers using telescopes for examining detail on the Moon or planets, many transparent nights being useless for this purpose. A brilliant night with rapidly twinkling stars is fine for the naked eye but not encouraging for work with a telescope. Good seeing is usually of little real benefit to the naked eye observer.

Transparent nights often occur in unsettled conditions with good seeing in settled hazy summer or misty winter conditions. The sky can be very clear after heavy rain or snow, between clouds on windy nights and when there is a dry frost. Transparency is more likely to be poor when there has been a daytime summer heat-haze, winter mist or hoarfrost. While weather forecasts give a very accurate general view of the weather in the British Isles, local weather can be so unpredictable that to rely solely on the forecast would mean missing some good observing opportunities. So the only certain way is to go outside and look. Even then it may cloud over ten minutes later.

It helps to get to know the local weather patterns in your area. Near the East Anglian coast the clearest nights are usually after a cold front has just been through with the wind in the north-west. Easterly winds in winter usually bring low cloud in off the sea just as you take your first look through the telescope on a hitherto beautifully clear evening. The cloud is often so low and thin that bright stars can be seen through it, but no useful observing can be done that night. In the summer, light southerly winds can be favourable. Weather systems can cross the British Isles in a few hours so if it is cloudy at dusk it may well be crystal clear before dawn. There are no safe rules except to look outside and hope for a clearance.

DARK SKIES

For seeing the fainter objects we need a dark sky. But often it is not very dark, even at night. There are two types of light sources preventing the sky from being really dark, natural and artificial.

Most of the natural sources have been discussed in previous sections and include moonlight, twilight, airglow and, less often, aurorae. We cannot do anything about them except to observe when possible in their absence. Of course if our object is only above the horizon in twilight or in moonlight, then we have to make the best of it.

Artificial sources include lights on the ground, street lights, floodlights, house lights, advertising lights, car lights and increasingly security lights. Sometimes the immediate effect can be reduced by finding a spot from which no light source can be seen. We cannot expect to see much in the sky if there is a bright light shining into our eyes as well.

More serious in urban areas is the form of light pollution that gives a bright glow above the ground level. From the middle of a large town this may cover the entire sky, often orange in colour because of the wide use of sodium lighting on roads. Even from country areas, the bright glow of a town can be seen for 60 km or more. But generally if you can get away from towns and main roads you will be able to see much fainter objects. The limiting magnitude quoted for the naked eye of six is only possible under a clear, dark sky and we cannot expect to see faint objects from a brightly-lit town centre.

The glow above a town or main road will be more troublesome on poor, hazy, misty or partly cloudy nights than on really clear nights, as the light we see has to be reflected down to us from particles or droplets in the atmosphere, either natural or artificial. Industrial haze and dust can add much to the glow and often depends on the wind direction in any one area.

CAMPAIGN FOR DARK SKIES

In several parts of the world a campaign has been started to try to halt and eventually reverse the ever increasing light pollution of the night sky. In the United Kingdom, the Campaign for Dark Skies (CfDS) was started by the British Astronomical Association (BAA) and is now combined with the Council for the Protection of Rural England (CPRE).

The aim is not to plunge the town and country back into the dark ages but to encourage well-shielded energy saving lighting for roads and public spaces and to encourage the use of security, house and industrial lighting that shines downwards and not upwards, wasting energy and preventing us and future generations from enjoying the night sky. Some limited progress has been made towards the use of better main road lighting and lighting for supermarket car parks, to take two examples.

OTHER SOURCES OF LIGHTS IN THE SKY

Aircraft lights are an increasing problem in most fairly densely populated countries. Even at distances of 30 km or more from an airport, lights as bright as a 0 magnitude star can usually be seen, often many at a time. An aircraft flying directly towards you at a distance of 20 km or more can appear as bright as Venus.

Dare it be said that artificial Earth satellites, although interesting 'astronomical' objects, are also a menace. It is now almost impossible to take a wide-field photograph of the sky requiring say 20–30 minutes exposure without having it spoilt by at least two or three satellite or aircraft trails and there can be many more. While aircraft can be avoided to some extent by keeping away from airports and the routes between them, artificial satellites respect no borders and may be photographed anywhere in the World.

DARK ADAPTATION OF THE EYES
Assuming that you have found a site with no light shining directly into your eyes and you want to see faint objects there are precautions you can take to improve your chances.

It is well known that the pupils of the eyes are small in bright daylight and become larger in low-light conditions. This change takes place quite quickly when we move from one to the other. But it also takes our eyes time to recover low-light sensitivity after exposure to bright light. It can take 20 minutes or more for full dark adaptation to be achieved. During this time even brief exposure to bright lights should be avoided. The need to become fully dark adapted will depend on the observing conditions. It is pointless on a night with a full Moon or when observing from a well lit area, but on a dark night it is very important. For consulting *The Times Night Sky* or other charts or books, use a small torch with a red glass or red paper cover as this has less effect on one's dark adaptation.

AVERTED VISION
While the centre of the retina is used for resolving fine detail and seeing colour, the peripheral areas of the eye are more sensitive to faint light. Astronomers use the trick of averted vision to see faint objects that may not be discernible when looked at directly. One looks a little to one side of where the faint object is and then one can often see it quite clearly. It is tempting to stare directly at it for a better look, when it may promptly disappear, as Alice found in the shop in 'Wool and Water', (Chapter 5 of *Alice Through The Looking Glass* by Lewis Carroll). With practice one can see more by averted vision when the object is near the limit of visibility.

HOURS OF DARKNESS
As a general rule one can take the time of the end of nautical twilight as the time when one can see clearly the brighter stars. On a clear, moonless night the Milky Way will probably not be visible until nearer the end of astronomical twilight, after which it gets no darker.

In this country Mercury has to be viewed in twilight and it helps to have a clear horizon free from obstructions. Bright comets, when they appear, may also be seen in twilight as they are usually brightest when near the Sun. Noctilucent clouds and artificial Earth satellites are visible only in twilight too. There is some advantage in observing Venus in daylight or twilight when binoculars or a telescope is used as this reduces the considerable glare, which can also be a problem with the Moon.

USE OF BINOCULARS
The use of binoculars will greatly increase the scope of what can be seen with the naked eye. Binoculars will often enable naked eye objects to be seen in less good conditions. Once the object has been identified it can often be seen with the naked eye.

Binoculars are described by the magnification and diameter of the aperture (of the large lenses) in millimetres, for example 7 x 50. The larger aperture gives a brighter image but also brightens the sky, so it may be better to use a 9 x 35 from brightly lit areas. Anything over nine times magnification is difficult to hold steady and of course binoculars of the larger apertures are heavier than smaller ones. These need to be supported on a substantial tripod.

DAYLIGHT OBSERVATIONS – A WARNING
It is dangerous to look directly at the Sun even with the naked eye and very dangerous with binoculars, a telescope or even a camera viewfinder. The result of a moment's lapse of concentration could be blindness.

Astronomers who study the Sun either project the image onto a white surface or use specially prepared filters. These are made by specialist firms and can be expensive. They should not be confused with those dark caps that fit over eyepieces on small telescopes. These may crack in the focused heat of the Sun and in any case do not protect the eyes at all necessary wavelengths.

If searching for Venus near the Sun with the eye or binoculars, always make sure that you cannot sweep over the Sun by mistake. Stand where the Sun is out of view, behind a wall or a thick tree.

OBSERVING VENUS IN DAYLIGHT
Apart from the Moon, Venus is the only object easily seen with the naked eye with the Sun above the horizon. Jupiter and Mars near their brightest are also possible but difficult, and it helps if they are near the Moon. This not only gives us direction but allows our eyes to focus for long distance also.

Venus can be followed in daylight in a clear blue sky. The best time is near greatest elongation when it is farthest from the Sun. It is easiest in the morning as it can be picked up at dawn and then followed through the day after the Sun has risen. The writer has followed it through using only the naked eye beyond midday when the Sun was in the south and Venus getting low in the south-west.

A few clouds can help one find the correct focus. Once Venus is found it is surprising how easy it is to see and find again. Its position can be fixed in relation to a tree or rooftop and bearing in mind that it will move westwards at about 15 degrees an hour, it is possible to estimate where it will be sometime later. In the evening pick it up when you can and having estimated its position, try earlier at the next opportunity.

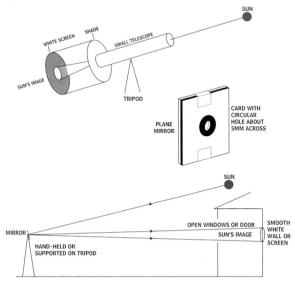

Fig. 9 *Projecting the Sun's image safely*

OBSERVING SUNSPOTS, SOLAR ECLIPSES AND TRANSITS

Without special equipment and specialist advice one should not observe any of these events through an optical instrument or directly with the naked eye. However, there are two safe ways to do so with minimal equipment.

If a telescope (binoculars can be used if less conveniently) is available, the enlarged image of the Sun can be focused on to a piece of white card; an image diameter of about 10 cm is useful. The card should be held or fixed some way behind the eyepiece of the telescope and the eyepiece adjusted until the correct focus is found. Another piece of cardboard with a hole in it to slide over the telescope tube will provide shade for the projection card. The aperture of the telescope need not be more than 50 mm. If any sunspots are visible they should be easily seen. Note that sunlight focused by the telescope can burn the skin and clothing.

The same set-up can be used for the partial phases of solar eclipses and transits of Mercury and Venus, when they occur.

There is a simpler way not requiring a telescope. A small plane (flat) mirror is used covered with a piece of opaque card in which a circular hole has been made about 5–10 mm across, the exact size is not important. The mirror is directed towards the Sun so that the Sun's image is projected on to a smooth white screen or wall. If the Sun is in the south a north facing room is best. Alternatively a large cardboard box could be used. The mirror can be supported on a camera tripod and will need turning slightly every ten minutes or so.

The image of the Sun will be about 12 cm across if the screen is 15 m from the mirror. The smaller the hole the sharper the image but the larger hole gives a brighter image. Two or three cards can be taped to the mirror, each with a different sized hole. Ensure that the smallest hole is put farthest from the mirror. Several people can watch an eclipse at once in safety and large sunspots can be seen. The image will not be not as sharp as that produced by a telescope. A back-silvered looking glass mirror will give a slight double image effect which can be avoided by using a front coated or metal mirror. But the image is sufficiently good for the purpose and provided one does not shine the beam into anyone's eyes and no one stares directly along it the method is safe. All demonstrations involving the Sun should be supervised by a responsible adult.

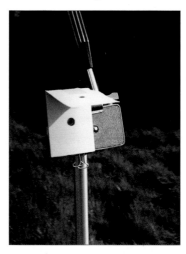

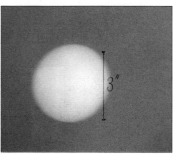

Top left: *A small looking glass mirror with two thin cards with different sized holes taped to one edge. The mirror is supported here on a camera tripod but it can, though less conveniently, be stood on a table or held in the hand. As the Sun moves, so the mirror must be tilted every few minutes to keep the image in view.*

Top right: *The mirror seen from behind projecting an image of the Sun through the open window on to the rear north-facing wall of a garden shed.*

Bottom left: *The projected image of the Sun. The distance between mirror and screen was about 10m (33ft) and the solar image about 90 mm (3.5 in) across. The 6 mm hole gave an image bright enough despite the shed having windows on the other three sides.*

A single lens reflex camera with standard 55 mm lens on a tripod. Note the lens hood and cable release. On modern fast colour films the stars and planets visible to the naked eye can be photographed in under 20 seconds using a fixed camera such as this.

SIMPLE PHOTOGRAPHY

Astronomers use cameras and telescopes that follow the westwards motion of the stars. But with modern films one can photograph some night sky phenomena with a fixed camera on a tripod.

While any camera with a time exposure facility and a fairly fast lens may be used, single lens reflex cameras are convenient and 35 mm film easily available and relatively cheap to use. Use a cable release to avoid vibration; one with a locking arrangement will hold the shutter open for as long as needed. The lens may dew up on cool damp nights. Keep it covered with a dry cloth when not in use and fit a lens hood or dew cap, easily made from cardboard.

The fashionable compact cameras where everything is electrically operated are not very suitable for this purpose because it may not be possible to keep the shutter open for long enough and if it can the battery may soon be exhausted. A manually-operated SLR camera is best. These are still available at a reasonable cost on the second hand market and usually have interchangeable lenses.

PHOTOGRAPHING STARS, COMETS AND AURORAE

For photographs of the stars and comets exposures should be kept short, to no more than 10–30 seconds, or trailing will be noticeable. The Aurora Borealis undergoes rapid changes so exposures should be less than a minute. Noctilucent Clouds can be photographed in 5–20 seconds.

So for all these phenomena a fast lens working at f/2 to f/4 focused for infinity on 200–400 ASA film should give reasonable results. There are faster films that will require less exposure still, though they may be more grainy. One has to experiment.

Black and white film can be used if you do your own processing. Colour print or slide film is easier to get processed: it is a good idea to take an ordinary picture on the first and last frames so that the processor does not cut up all your films (which may appear to have nothing on them) in the wrong place!

For photographing star trails exposures up to several hours can be made but at smaller aperture (f/16) and on a slower film (ASA 100) or the film will be fogged. On colour films the trails show the stars different colours. An exposure on stars near the celestial pole will show the stars as circles. A wide angle lens (e.g. 30 mm focal length) will cover more sky.

PHOTOGRAPHING METEORS

For photographing meteors, a fixed camera may be used. A fast lens of about f/2 and a fast film of 400 ASA or faster is best. The camera will need a bulb or time exposure. How long the exposure can be will depend on the brightness of the sky, but five minutes or more can be tried. Fix the camera to a tripod at an angle of about 50 degrees to the horizontal and about 60 degrees from the meteor radiant. A series of exposures can then be made.

Even during a strong shower such as the Geminids the number of meteors photographed will be disappointingly small. Most will miss the area covered by the camera and many will be too faint, even though appearing quite bright to the naked eye. However, on the night of maximum on a roll of film one might well record several meteors, as well as artificial satellites and of course aircraft lights. The longer the total exposure time, the more meteors should be recorded. If a meteor storm should develop during the Leonids in the next few years, those in the right place at the time could record a few dozen on one exposure, as happened in the USA in 1966.

While a wide angle lens will cover more sky, if it is of smaller aperture it will miss fainter meteors that an f/2 might record. Because a meteor is essentially a (moving) point image, larger apertures photograph fainter meteors.

PHOTOGRAPHING ARTIFICIAL SATELLITES

The technique for photographing artificial Earth satellites is similar to that for meteors. If a prediction for the satellite is available, the camera can be pointed before the satellite appears and the shutter opened for as long as it takes it to cross the field of view.

PHOTOGRAPHING CONJUNCTIONS

For conjunctions of planets with each other, bright stars or the Moon, a telephoto lens may be used if it includes all the objects. Those in twilight make attractive pictures on colour film but exposures will need to be kept fairly short if a fast lens and film is used, just a few seconds. Try several exposures, make notes, and try to avoid making the same mistake next time. This is how we all learn to judge, sometimes successfully, what is needed for the next occasion.

RARE OR UNUSUAL OBSERVATIONS

Over the years I have been fortunate enough to have witnessed a number of unusual astronomical phenomena. Some of these were rare or even very rare. Others may happen every few years but the opportunities to witness them are relatively few, requiring the event to take place at a time when it is visible from your particular location on Earth, clear skies, perhaps the right observing equipment and lastly but by no means least, being free and able to go out and look at the appointed time. Astronomical phenomena wait for no one, not even astronomers. To miss something by only a second is no consolation: even blinking at the wrong moment can sometimes make the difference between success and failure.

Some phenomena, like a very bright meteor (fireball or bolide), just happen without warning; one cannot plan to see one, but those looking at the sky more often, especially during showers like the Leonids, stand a better chance of doing so. Others, like eclipses of the Sun or Moon, can be forecast precisely years or even centuries ahead. No special equipment is needed just to see them, but you have to be in the right place at the right time and this usually means travelling to another part of the globe. Somewhere in between come, for example, bright comets. Their position in the night sky can often be predicted weeks or months ahead and they can often be seen for weeks from a whole hemisphere but no one can be certain how bright or interesting they will be.

There is often disappointment where comets are concerned. Astronomers have a problem. Do they announce a forthcoming event and risk disappointment or criticism if it transpires that there is nothing much to see. Or do they say nothing and risk depriving the general public of an opportunity to see something unusual that they may remember all their lives? I hate to miss seeing something interesting so if there is any chance at all I try to have a look. Sometimes the clouds break at the crucial moment and predictions can be unduly pessimistic as well as too optimistic. The only way not to miss anything is to go out and have a look for oneself. To do this one has to know when something is likely to be observable. It is this information that the annual *The Times Night Sky* booklet and monthly notes in *The Times* seek to provide.

A few of my more memorable observations are described below. Several were made possible by last minute improvements in the weather, some were just more visible than predicted and a few were not predicted at all and seen just by chance. In a climate like that of the British Isles, where it can be difficult to predict the state of the sky even a few minutes ahead, one must expect to be frustrated by the weather as often as not.

CONJUNCTION OF URANUS AND JUPITER

During May 1955 Jupiter passed almost in front of Uranus, so that for a while they were very close, only 10 arcminutes apart on 10 May. Uranus is about 5.7 magnitude and comparable in brightness to the four great satellites of Jupiter. About this time one could see Jupiter with an apparent fifth bright satellite. As Jupiter takes about 12 years to move right round the sky and Uranus as long as 84 years, Jupiter will pass Uranus at intervals of about 14 years, but they are seldom as close as in 1955, and some conjunctions take place when the planets are too near the Sun to be observable.

OCCULTATION OF MARS BY THE MOON

This occultation of a planet by the Moon was seen from the middle of London Bridge in the city of London at eight minutes to midnight on 6 February 1957. The Moon was about 5 degrees above the western skyline. Mars was visible to the naked eye as the crescent Moon edged towards it. In binoculars Mars faded over three or four seconds before disappearing behind the Moon's dark limb. By the time of reappearance at the bright limb the Moon was too low for observation.

AURORA OF 28 OCTOBER 1961

This aurora was first noticed from a (then) very dark location in northeast Essex at 17.30 UT as an unusual brightness along the north-west horizon when it should have been almost dark. About 18.00 a narrow band of light was visible moving from ENE to WSW, later doubling along its length. Groups of rays appeared above the N-NW horizon, together with patches of light varying in brightness rapidly. At times these rays reached to the zenith and beyond. A reddish glow was confined to the western end of the display. At 18.30 the whole of the northern part of the sky to the zenith was white, like moonlight on clouds. The brighter stars of the Plough and Auriga were almost invisible and all fainter stars vanished in the glow. An isolated arch 5 degrees wide formed from ENE to WSW just south of the zenith and persisted for about ten minutes. It was much brighter than the Milky Way. Arches one above the other were visible in the north with darker sky in between. A rapidly formed arch moved quickly away from the north. This was repeated several times. Patches of sky glowed momentarily brighter like a breeze blowing over hot coals. The general activity shrank back towards the northern horizon and by 19.30 the sky was almost back to normal with the Milky Way again prominent. The Moon rose at 20.30 UT.

Spectacular displays are commonplace in many parts of the world with some activity seen almost every dark night. Red, green and yellow as well as white are often prominent. These displays occur mostly within a few thousand kilometres of the geomagnetic poles. Even in southern England the aurora borealis can be exciting though great displays may be decades rather than years apart. They tend to occur around the time of maximum solar sunspot activity and are often predominantly white with areas of red or perhaps green. The spread of light pollution has made it less likely that such a display would be noticed from urban areas.

21 December 1968 at 18.20 UT. Fuel dumped from an upper stage of the Saturn V rocket glowing in sunlight. The Apollo 8 spacecraft, which could be seen as a bright star through the telescope ahead of the cloud, was some 50,000 km from Earth when this photograph was taken from north-east Essex. The three U.S. astronauts were the first to reach the Moon and were in orbit round it over Christmas 1968. Apollo 11 landed the first men on the Moon in July 1969.

APOLLO 8

On 21 December 1968 Apollo 8 lifted off from Cape Kennedy (now Cape Canaveral again) carrying three astronauts (Borman, Anders and Lovell) on a mission to fly within about 100 km of the far side of the Moon as part of the preparation for a first lunar landing, successfully accomplished the following July by the crew of Apollo 11.

That evening, a Saturday, the western sky was clear and Venus was brilliant in the SSW. At about 16.45 in strong twilight a hazy star was seen some degrees below Altair. Through the telescope the object showed a small disc surrounded by a hazy glow. It was soon clear that the object was moving relative to the stars, so could not be a nova seen through cloud or a comet, and its motion was much too slow to be an aircraft or artificial satellite. After telephoning another astronomer and coming outside again, the object had become a small diffuse bright disc about half the size of the Moon. Some photographs were taken by which time the object had become faint and nearly 3 degrees across. By 17.35 it was no longer visible. My supposition was that it was connected with Apollo 8.

On going out again at 18.06 I was surprised to see a bright hazy star of about 0 magnitude, now to the east of Altair. A bright starlike object was visible in the telescope ahead of the glow, believed to be reflected sunlight from the Command and Service Module itself. By 19.00 it was about the size of the Moon but crescent shaped, fading to invisibility by 19.20. There were no further outbursts. It was confirmed the next day by telephoning a colleague at the Royal Greenwich Observatory that the objects were on the predicted path of Apollo 8.

It was later established that surplus fuel had been vented and that the Command and Service Modules were about 50,000 km from the Earth. Photographs taken from Spain at about the same time showed the parallax effect for an object near the Earth which was consistent with this order of distance. Although similar events must have taken place on subsequent missions, the timing and circumstances prevented a repeat of these unusual observations so far as I was concerned.

TRANSIT OF MERCURY
Every so often Mercury passes between the Earth and the Sun and it can then be seen as a small black disc, not unlike a small sunspot, crossing the Sun's face (see p.73). Not all events can be observed at any one place. The 9 May 1970 transit was observable from the British Isles and it was also Saturday, but thick low cloud was reluctant to break up. Preparations had been made to project a 20 cm image of the Sun using a 13 mm eyepiece on a 15 cm reflecting telescope: the telescope was driven to follow the Sun, which could be seen from time to time as a blur behind the clouds. All this was very frustrating as the whole transit lasted over seven hours. Eventually about 12.10 the Sun broke through momentarily and a small black notch was seen in the Sun's limb. This was Mercury just leaving the Sun's disc. Of course the rest of the day was mainly sunny!

OCCULTATIONS OF SATURN BY THE MOON
Sunday 3 February 1974 was a fine, clear day. Venus was followed in my 125 mm f/17 refractor from 10.00 until well past the meridian at 13.00. There was to be an occultation of Saturn by the Moon around 15.00 which I thought might be observable. Saturn was not seen before its disappearance at the Moon's dark (eastern) limb but a short while after the time of reappearance at the western limb Saturn was seen about a ring's diameter from the Moon. It was followed in the telescope until after dark. In daylight, the Moon's bright limb does not look very bright, being a pale white colour against the blue sky. Saturn was even fainter, a mere ghost with globe and rings looking very small alongside the large curved limb of the Moon. Saturn is not easy to see in sunlight and being near the Moon helped one know where to look and gave something to focus the eyes upon.

Occultations often happen in series and at the next lunation Saturn was again occulted, this time about midnight on 2–3 March. An accurate timing of the disappearances of the planet behind the Moon was made. Saturn's brightest satellite Titan (8.3 mag.) also disappeared and was timed. The reappearance of Saturn nearly an hour later was also timed but thin cloud hid Titan until it was well clear of the limb.

OCCULTATION OF A STAR BY SATURN'S SATELLITE TITAN

A very rare event took place on the evening of 3 July 1989 when Saturn's largest satellite Titan (8th magnitude) passed in front of the 5.4 magnitude star 28 Sagittarii. No occultation of a star by Titan had ever been observed and it could be hundreds of years before it happens again. Titan is tiny as seen from the Earth and there was some uncertainty as to the precise track on the Earth's surface from which it would be seen. As the time of disappearance approached, Titan merged with the star as seen in the telescope. The event itself was signalled by a drop of about 2 magnitudes in the brightness of the star. Incredibly it was clear over a large part of western Europe and the occultation was widely observed. Despite Saturn's rather low altitude I was able to observe and time the whole event using a 25 cm reflector. It was a unique experience.

JUPITER AND COMET SHOEMAKER–LEVY 9

One of the most exciting series of observations concerned the nuclei of comet Shoemaker-Levy 9 and Jupiter. It is beyond the scope of these notes to describe the progress of the whole event, which has been the subject of several books, but the comet's nucleus had already split in to at least 22 separate pieces which were predicted to impact on the far side of Jupiter from 16 July 1994. There were conflicting opinions as to whether anything would be visible from Earth, let alone with amateur instruments.

From the British Isles Jupiter was not well placed and by the time it was beginning to get dark it was low in the south-west and at times seen amongst the branches of a tree. The seeing was generally bad with patchy cloud and haze. I was not expecting to see anything unusual when I turned my 15 cm refractor towards Jupiter at about 21.20 on 18 July. There was much passing cloud and the image was often blurred due to the unsteady atmosphere. Even so, two new dark patches in the southern hemisphere were immediately apparent. The positions could be related to earlier impacts of the comet's nuclei. Further observations were made on July 19, 20, 21, 22, 23, 26, 29 and 1 August, this last showing six dark areas. Although observing conditions were never good, the new impact sites were unmistakable, and most could eventually be related to specific nuclei as determined by professional astronomers using large telescopes on mountain top sites.

It was the first time that the bombardment of another planet by a comet or asteroid had been observed, although a look at the Moon's impact craters shows that it must have happened thousands of times in the past. That it could be seen in telescopes as small as 5 cm aperture was quite unexpected and it enabled thousands of astronomers both amateur and professional to witness this unique event.

ALL THE PLANETS VISIBLE IN ONE DAY

On 2 October 1996 all the planets, except Pluto, were seen within 12 hours. In morning twilight Mercury was quite bright in the east, with brilliant Venus near Regulus with Mars also visible, all to the naked eye. Mercury and Venus were also seen during the day in the telescope. In the early evening Jupiter and Saturn were visible to the unaided eye and Uranus and Neptune were seen in the telescope. As a bonus the Moon and comet Hale-Bopp were also seen. Opportunities to see all the planets on one date are not exactly rare, I have seen them all several times. But it often happens that one planet is too near the Sun to be observed when all the others are available.

THE 1998 LEONIDS METEOR SHOWER

The orbit of comet Tempel-Tuttle passes close to the Earth and particles left in its wake give rise to a meteor shower that reaches peak activity every 33 years. Usually increased activity recurs on about November 18 for several years running but numbers of the fast-moving meteors seen vary from a few per hour to tens of thousands.

With the comet nearest the Sun in 1998 increased activity was expected over 1996–2001. Peak rates reached more than 200 per hour, but on the morning of 17 November observers in Europe were treated to a display of bright fireballs some hours before the peak was predicted. In a moonless sky the display was outstanding with many beautiful fireballs often ending in bursts brighter than Venus and some bright enough to cast shadows. Some meteors left glowing trains lasting several minutes which rapidly distorted in the strong upper atmospheric winds. With over 230 Leonids seen in about four hours despite the need to observe from indoors on this occasion, it was a night to remember, if not a meteor storm such as the Leonids have given in the past. It was quite the most exciting meteor display I have seen in over half a century of observing meteors.

NOTES ON BINOCULARS AND TELESCOPES

USING OPTICAL AID

The Times Night Sky is a guide for observers using the unaided eye but as many readers will have available a pair of binoculars mention is made of how they can enhance their enjoyment of the heavens. While binoculars can give fine low power views of the night sky, the magnification is too low to show any detail on planets or other small objects. Where higher magnifications must be used or where more light must be collected, an astronomical telescope is needed. Good telescopes are expensive but they have a long life. Brief notes on what to look for in buying a first telescope are also appended.

BINOCULARS – TYPES SUITABLE FOR ASTRONOMY

As was explained in the previous section, binoculars are rated by the aperture in millimetres of the light gathering lenses called objectives (sometimes object glass or OG) and the magnification, which is a linear measurement not one of area. For example a 9 x 35 binocular has objectives with an unobstructed aperture of 35 millimetres and will make the distance between two stars (or two trees) appear nine times greater than seen with the naked eye.

Larger aperture objectives will collect more light and show fainter stars, but if they are made larger without increasing the magnification, there soon comes a point when all the light collected will not enter the eye, and the additional aperture is wasted. For this reason large binoculars with 80 mm objectives are made with relatively high, 15 x or 20 x, magnifications. Larger binoculars with higher magnifications show less sky at one time (have a smaller field of view) and this can make objects more difficult to find, though objects appear bigger and stars brighter. The magnification usually needs to be higher for older observers and those observing from a bright site because the eyes' pupils dilate less when we are older and in brighter conditions.

Experienced observers have special needs, but for an extension of naked eye observing a smaller, lighter binocular is to be preferred. The 7 x 50 used to be the standard recommended for astronomy but the brighter skies many of us have to contend with may make 9 x 35 or 10 x 40 a better choice. The weight of binoculars goes up quickly as the aperture increases. What may be an acceptable weight for looking horizontally with the elbows braced against the chest may soon become tiring when trying to hold a point of light in view when it is above ones head.

Anything over 13 x 60 certainly needs support and I find even this size difficult to hold steady for more than a few seconds. The latest thing in binoculars is a self-stabilising mechanism that holds the image steady even if the binoculars themselves wobble or shake. At present these are very much more expensive than the ordinary type.

BUYING BINOCULARS

If you are buying for the first time a pair of general purpose binoculars with astronomy in mind, my advice would be not to go for too large a pair, to buy from a reliable source, preferably where they know about astronomy and telescopes, and to err on the side of quality if you can afford it. Mechanical robustness is important for an instrument that will be used in the dark and probably be sat on from time to time. A good pair of binoculars will last a lifetime: my most-used 9 x 35s are 40 years old and as good as new. Modern binoculars have more and better lens coatings on their lenses but otherwise differ little from the older types. If possible take along the owner of a pair you know are satisfactory at night and compare them to what you are proposing to buy.

Test your proposed purchase at night if possible but if not, on a distant view, which may have to be along the street. They should not show distortion at the edges of the field, that is to say a telephone pole or vertical edge of a building near the left or right edges of the field of view should be straight and not strongly bowed inwards (pincushion distortion) or outwards (barrel distortion). It should not be strongly coloured along its edges. The specular reflection of the Sun from a distant object can sometimes be used as an artificial star. A star should appear small, sharp and when put slightly out of focus, round. The image should be comparable in appearance near the edge of the field, though perhaps not quite as good as at the centre where critical observations will be made.

EYE DEFECTS

One must be a little careful about rejecting optical instruments when it may be ones own eyes that are at fault. Most of us have some eye defects, for example, seeing rays around a bright star or Venus; these tend to get more noticeable with increasing age. They will probably be seen in binoculars too. If you wear spectacles for certain eye defects, you may need to wear them with binoculars or a telescope. Try this out before buying. Many binoculars have rubber eyecaps that will prevent scratching of the lenses and sufficient eye relief for spectacle wearers. It is important that one can get the eyes close enough and still see the whole field of view.

MOUNTING BINOCULARS

Many binoculars have a standard camera tripod screw thread set into the body which makes attachment to a camera tripod simple. There are other gadgets and means of supporting them but this may involve buying or making something extra. It is possible to buy a monocular which is essentially one half of a binocular. Many prefer to use both eyes but generally with a telescope one uses only one. Monoculars can be considerably cheaper for the same quality and are much lighter in weight and are worth considering.

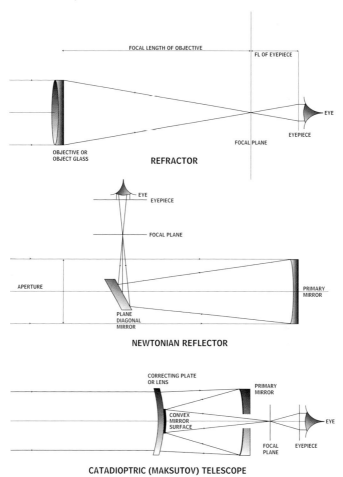

Fig. 11 *Telescope types*

TELESCOPES – ASTRONOMICAL TYPES

Much of what has been said about binoculars also applies to small telescopes. A monocular is a small refracting telescope which consists basically of an objective and an eyepiece. But in monoculars, binoculars and spotting telescopes used by bird-watchers and others an arrangement of prisms is used to make them more compact and turn the image the right way up. A simple refractor inverts the image, things appear upside down and what is to the left appears on the right, which can make finding things on the sky tricky until one gets used to it, but one soon does. Astronomers prefer this to the mechanical and optical complication and loss of light (now quite small with modern lens and prism coatings) that re-inverting the image involves.

BUYING A TELESCOPE
It is difficult to give advice on acquiring a first telescope. Much depends on what use is likely to be made of it and how much one is willing to spend. Quality telescopes are not cheap, but it is false economy to buy one of poor design and quality. A good telescope will last a lifetime. The one I use now was built in 1914 and before that I used one built in 1880, now in regular use by another astronomer. It is beyond the scope of these notes to describe telescopes in detail. There are good books on the subject. Many local and national astronomical societies will be keen to advise. If at all possible get a little telescopic experience and advice before buying.

One can only advise against buying the small telescopes sometimes seen in high street stores. Some may be suitable for young beginners but many are not and anyone with a long term interest in astronomy is unlikely to be satisfied with one for very long. Some poor examples have simple objectives, stopped down inside to a smaller diameter to improve the image but with much loss of light. The magnifications quoted are usually quite unrealistically high and the eyepiece is often of poor quality. The simple table stands with hinge joints are unlikely to allow smooth movement and in any case you could only use them for objects not far from the horizon.

If you have to go it alone, go to a reliable astronomical instrument dealer; the names and addresses of some can be found in some monthly astronomical publications. Obtain their catalogues before visiting them and take your time in deciding; you may be using the same instrument for the rest of your life. The choice is likely to be between a refractor of about 80 mm (3 inches) aperture or a Newtonian reflector of about 150 mm (6 in) aperture. One might expect the cost to be about the same, several hundred pounds.

REFRACTORS VERSUS REFLECTORS
The difference in the performance of the two types is not so very great: the larger aperture will show slightly fainter stars and allow slightly higher magnifications to be used. Reflectors need readjusting from time to time and the mirrors will need a fresh coat of aluminium, but with care only every few years. Refractors need no servicing and being smaller are easier

to transport, set up and mount. There are now some small catadioptric (lens and mirror) instruments worth considering. These are very compact but may not be as useful in the long run as a simple telescope.

MAGNIFICATION AND EYEPIECES

The standard size of eyepiece (the diameter of the lens barrel) is 1.25 inches (about 32 mm). Some small telescopes use 24.5 mm size. These are best avoided. There are far more eyepieces (called oculars in the USA) and accessories available for the larger standard size and these can always be used if you change to a larger telescope later. Telescopes are often supplied with one or more eyepieces, usually a 26 mm focal length which would give a magnification of around 40 x on an 80 mm, a f/13 refractor or 50 x on a f/8 reflector. Other useful powers would be 13 mm (80 x, 100 x) and perhaps 7 mm (150 x, 180 x). As a general rule 150 x would be about the highest magnification or power one would use on an 80 mm telescope. On rare occasions one might be able to use 300 x on a 150 mm telescope. The higher the magnification the smaller the area of sky seen at one time, that is the field of view is smaller.

MOUNTING THE TELESCOPE

Astronomical objects are continually moving across the sky and to be of any use the telescope must be able to follow this movement smoothly. There are two main ways of mounting a simple telescope, the alt-azimuth and the equatorial. With the alt-azimuth the telescope is mounted like a gun. It turns along the horizon around a vertical axis, and can be elevated or lowered about a horizontal axis. Both must move freely and be capable of very small adjustments. To follow an object on Earth you would just move it about its vertical axis. To follow a star or planet you must do this and raise or lower it a bit about its horizontal axis at the same time. This is quite easily done in practice if the mounting is well made. For reflectors the Dobsonian form of the alt-azimuth mounting, usually made of wood, can be all that is needed but it is not suitable for refractors.

The equatorial is in effect an alt-azimuth that has been tilted until the vertical axis points north and south, parallel with the Earth's axis. Then a single movement about this polar axis enables an object to be kept in view. In many telescopes this axis is driven by some kind of electric motor drive (often called a driving clock from the time when most were weight driven) so that following an object is continuous and automatic. Computer control can now be used to make suitably equipped alt-azimuth telescopes follow objects automatically in the same way, but such telescopes are relatively very expensive.

A strong, sturdy tripod or pillar is needed to support a telescope. The type of mounting is less important than its ability to support the instrument without vibration and allow it to move smoothly. Photographic exposures longer than a few seconds require an equatorial though.

Above left: *A small wide-field 125 mm aperture refractor mounted in trunnions on a tripod, the simplest type of alt-azimuth mounting. For higher magnifications a means of moving the telescope smoothly in altitude and preferably also in azimuth is desirable (these controls are known as 'slow motions').*
Above right: *A home-made 150 mm aperture f/8 Newtonian reflector carried on a simple alt-azimuth fork mounting. The mounting was in three parts allowing easy transport by car.*

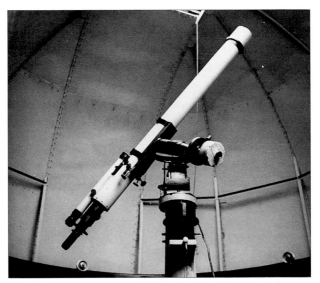

A 110 mm refractor mounted on a clock driven equatorial. Mounting a telescope on a fixed pier in an observatory keeps it permanently aligned on the pole and meridian and ready for immediate use. It also protects it and the observer from stray light, wind and dew when in use.

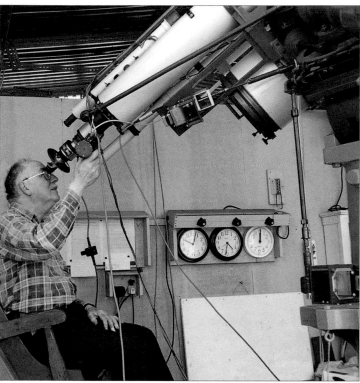

The author's 150 mm refractor weighs over 500 kg (half a ton) and stands on a cubic metre of concrete. In the shorter fatter tube is a 150 mm aperture camera lens for photographing comets and it carries two smaller refractors. Stepper motors drive the telescope in Right Ascension at the sidereal or solar rate and allow precise adjustments in RA and declination. The observing chair can be raised or lowered electrically as the position of the eyepiece changes. The pitched roof is in two halves which roll away on rails. The framework of the building is of slotted steel angle and the roof and outside walls of corrugated PVC. The materials used adjust quickly to the ambient temperature and the open design allows rapid circulation of the air, both of which can improve the quality of the seeing.

PORTABILITY AND CONVENIENCE

In a climate with rapidly changing weather conditions, convenience is very important. Many serious amateur astronomers build their own observatories so that their telescopes are always ready to use with minimum delay. This involves quite a large investment in time and money. For telescopes that are going to be kept indoors and taken out each time it looks clear, portability and ease of setting up become very important. For this reason refractors larger than 80–100 mm and reflectors larger than 150–200 mm can hardly be considered portable.

If a telescope takes a long time to take outside and get ready, it will not be used very often. A smaller, lighter instrument will be much more convenient and will show nearly as much anyway and being used more often you will actually see more than with a larger one.

WHY BUY A TELESCOPE?

What can you see that you could not see with binoculars ? You can see an amazing amount of detail on the Moon. As the Sun rises over the lunar landscape, the shadows can be seen retreating, valleys filling with light. It is an ever changing show. The phases of Venus can be followed, markings on Mars are rather vague except when it is near opposition, Jupiter's rapid rotation brings fresh areas of its cloud belts into view in only a few minutes and the constant motions of the four great satellites can be followed as they are occulted, eclipsed or their shadows cross the planet. Saturn's rings and brighter satellites can be seen and the tiny discs of Uranus and Neptune. There are many double stars, variable stars, nebulae and galaxies to see. Comets and minor planets can be followed.

A telescope opens up a whole new world in much the same way as binoculars do over observing with only the unaided eye. It is best to work up in steps. Know your way about the sky with only the eye, add binoculars, then perhaps make the giant leap and acquire a small telescope. All later steps upwards in telescope size will be something of a disappointment by comparison with one's first views through a good small telescope.

FURTHER READING

Astronomy from Towns and Suburbs, Robin Scagell, George Philip Ltd., 1994.
The Aurora: Sun-Earth Interactions, Neil Bone, Ellis Horwood, 1991.
Guide to the Sun, K.J.H. Phillips, Cambridge University Press, 1992.
Meteors, Neil Bone, George Philip Ltd., 1993.
The New Solar System, Fourth Edition, Ed. Beatty, Petersen & Chaikin, Cambridge University Press (and Sky Publishing USA), 1999.
Norton's Star Atlas, Nineteenth Edition, Ed. Ian Ridpath, Addison Wesley Longman Ltd., 1998.
Year Book of Astronomy, Ed. Patrick Moore, Sidgwick & Jackson Ltd., (each year).

ASTRONOMICAL SOCIETIES

Society for Popular Astronomy (formerly Junior Astronomical Society), The Secretary, 36 Fairway, Keyworth, Nottingham, NG12 5DU.
The British Astronomical Association, The Assistant Secretary, Burlington House, Piccadilly, London WIV 9AG.
The Federation of Astronomical Societies, FAS Publications Secretary, Tabor House, Norwich Road, Mulbarton, Norfolk, NR14 8JT. (This organisation can also advise on who to contact in many local and regional astronomical societies.)

APPENDIX: PHOTOGRAPHIC DETAILS

The copyright in the photographs and illustrations belongs to Michael Hendrie, except where otherwise stated.

p.29 Milky Way in Scutum. Dallmeyer /2.5 portrait lens of 215mm focal length, 40 minutes on a Kodak O.800 plate, taken in 1956 from amongst the street lights of Westcliff-on-Sea, Essex. Milky Way in Cygnus. Cooke f/4.5 lens of 460mm focal length, 20 minutes exposure on an Ilford Zenith Astronomical plate.

p.31 The Great Galaxy in Andromeda. Exposure of 7 minutes covering 9 x 7 degrees of sky. Dallmeyer f/2.5 lens of 215mm focal length on a Kodak O.800 plate.

p.33 Pleiades. Cooke f/4.5 lens of 660mm focal length, 15 minutes on a Kodak IIa-o plate. Praesepe. Ex-RAF reconnaissance lens f/5.6 of 355mm focal length, 15 minutes on a Kodak O.800 plate.

p.34 Orion, main picture. Wray f/4.5 wide angle aerial survey lens of 300mm focal length, 114 minutes exposure on a Kodak Oa-J plate. Inset. Reflector f/7 of 1780mm focal length, 10 minutes exposure on a Kodak IIa-o plate.

pp.38, 39 Nova Cygni. Zeiss f/7 lens of 1200mm focal length. Exposures of 2 and 5 minutes on Kodak TMax 400ASA film. Photo: H.B. Ridley.

p.55 Moon first quarter. 1/15th second on Ilford Delta 100 ASA film. Moon gibbous. 1/125th second on Ilford HP5 400 ASA film. Cooke f/13 refractor of 1950mm focal length with a yellow filter.

p.59 Moon and Venus. Kodak Aero Ektar of 180mm focal length. The f/2.5 lens was stopped down to f/6 to improve definition at the edges of the picture. Exposure 1 second on Kodak TMax 400 ASA film. Photo: H.B. Ridley.

p.62 Total eclipse of the Sun. 400mm focal length f/5 lens, 1 second exposure on 200 ASA film.

p.64 Partial solar eclipse: Cooke f/13 refractor of 1950mm focal length with a Thousand Oaks glass filter (density 5) over the object glass. Pentax Spotmatic camera body with lens removed, 1/30th second on York Photo 100 ASA colour print film.

p.65 Total eclipse of the Moon. Partial phases, f/8 reflector of 1200mm focal length stopped down to f/12, 1 second exposures on Ilford HP3 400 ASA film. Totality, ex-RAF f/5.6 lens of 355mm focal length, 3 minutes exposure on an Ilford Astra III plate.

p.74 Venus and Jupiter. Pentax Takumar f/1.8 lens at f/2.8, 55mm focal length, 4 seconds on Kodacolor 100 ASA film. Moon, Venus and Saturn. Pentax Takumar f/1.8 lens of 55mm focal length, 7 seconds on Kodak VR 100 ASA film. Moon and Jupiter. Pentax Takumar f/3.5 lens of 135mm focal length, 5 seconds on York Photo 100 ASA film.

p.75 Venus and Jupiter. Pentax Takumar f/3.5 lens of 135mm focal length, 5 seconds on Kodak Gold 200 ASA film.

p.82 Uranus and Neptune. Cooke f/4.5 lens of 660mm focal length, 10 minutes exposure through an orange filter on Kodak TMax 400 ASA film.

p.83 Comet Hale-Bopp. Pentax Takumar f/3.5 telephoto lens of 135mm focal length 15 minutes exposure on Fujicolor Super G Plus 400 ASA colour print film. Comet Hale-Bopp. Cooke f/4.5 lens of 660mm focal length, 20 minutes on a Kodak IIa-o plate.

p.85 Eros. Dallmeyer Serrac f/6.3 lens of 355mm focal length, 40 minutes on an Ilford Selo plate. Photo: H.B. Ridley.

p.89 Leonid meteor. Pentacon f/1.8 lens of 50mm focal length, 8 minutes on Ilford HP5 400 ASA film.

p.93 Noctilucent clouds (1992). Pentax Takumar f/1.8 lens of 55mm focal length, 10 and 20 seconds exposures on Kodak TMax 100 ASA film. Noctilucent clouds (1966). Zeiss Tessar f/2.8 lens of 80mm focal length, 1 second on Ilford FP3 125 ASA film.

p.95 Sputnik 2. Dallmeyer f/2.5 portrait lens of 215mm focal length, 20 seconds on an HPs plate.

p.97 Contrails. Pentacon f/3.5 wide angle lens 30mm focal length.

p.109 Apollo 8. Zeiss Tessar f/2.8 of 80mm focal length, 3 minutes exposure on Kodak Tri-X 400 ASA film.

INDEX